D1154820

COLLECTIONS OF POEMS
BY RON LOEWINSOHN

Watermelons
L'Autre
The Step
Meat Air
Goat Dances

EDITED BY RON LOEWINSOHN

The Embodiment of Knowledge,
by William Carlos Williams

MAGNETIC FIELD(S)

RON LOEWINSOHN

MAGNETIC FIELD(S)

A NOVEL

MIDDLEBURY COLLEGE LIBRARY

Alfred A. Knopf New York 1983

PS
3523
.032
M3
1983

7/1983
gen'l

THIS IS A BORZOI BOOK
PUBLISHED BY ALFRED A. KNOPF, INC.

Copyright © 1983 by Ron Loewinsohn

All rights reserved under International and Pan-American Copyright Conventions. Published in the United States by Alfred A. Knopf, Inc., New York, and simultaneously in Canada by Random House of Canada Limited, Toronto. Distributed by Random House, Inc., New York.

This is a work of fiction. All of its characters and all of its incidents are imaginary, made only of words. If any of them resemble characters or incidents in what we call the real world, that resemblance is purely accidental.

Library of Congress Cataloging in Publication Data
Loewinsohn, Ron.
Magnetic field(s).
I. Title.
PS3523.O32M3 1983 813'.54 82-48879
ISBN 0-394-53105-1

Manufactured in the United States of America
First Edition

For Stephen, Will, and Joe, and
for Kitty, who brought me cups of coffee
all that dark autumn in Reykjavik

PART ONE Albert 3

PART TWO Kindertotenlieder 45

PART THREE Daniel 135

MAGNETIC FIELD(S)

1 / ALBERT

Killing the animals was the hard part. "All you've got to do," Jerome had told him, "is keep your cool. There isn't anybody else's cool you've got to keep, and there isn't anybody else who will keep yours." The first time it happened—he had already gone with Jerome five or six times—they were in this backyard when this little black dog (a Scottie?) started yapping up at them and he froze in the first moment and in the next felt his body want to move back over the fence, the way they'd come. But Jerome had simply reached down to pick up a brick that formed part of a kind of border around some flowers—it was all one smooth motion—and bashed in the dog's head with one good one, and then gave it one more to make sure, the second one making an awful sound as it went into what was left of the dog's head. There were just these four little legs attached to this furry black body, and where the head had been was just this turmoil of hair and blood and meat and one piece of jagged bone and one eye where it shouldn't have been, not even not looking at anything, just there; and he looking down at it till Jerome touched him on the chest and then motioned toward the window with one jerk of his thumb.

The first time he'd had to do it himself, they had just stepped into this backyard at the end of the driveway and suddenly there was this big setter—just as surprised as they were. He started to feel himself wanting to turn back out of there, hearing the blood rushing in his ears and thinking he did not want to do this as he reached back, his hand closing around the cold shaft of the tire iron in his back pocket. Behind him Jerome had stopped. He reached out his left hand, offering "something" to the dog, which leaned forward to check it out at the same time that it began to pull its lips back away from its teeth. He brought the tire iron out and down in one sweeping arc into the animal's head. It actually made a dent in the head and felt something, at first, like hitting a rolled-up rug, except that now there was this dog lying there with its different legs crumpled under it or sticking out in a way that nothing that was alive would lie that way. He just stood there a long moment, looking down at that furry body, hearing himself breathing and hearing the dog not.

Killing the animals was the hard part because they tended to get under his skin; they reminded him of him. Their bodies were complicatedly thick, so that hitting this skull was nothing at all like hitting a rolled-up rug, or a log, things that were solid, the same all the way through. They were made of all those layers of skin and bone and organs, all of which he thought of as easily bruised, especially the organs. They moved like him, they wanted things, they wanted, and they were scared and furious.

They were also absolutely different. That's how he could hate them this swiftly, this coolly. They were part of this place. They defended it dumbly, the way a door would just stand there keeping people out till somebody kicked it in or bashed it in with his shoulder or an axe. They lived in this place like the kid's bike that Jerome was now stepping over on his way to the back steps. They belonged to something and they never even thought about it, not even when they fought or even died to protect it as if it were their life. Getting inside the house itself was easy.

Jerome first of all reached up to the lintel above the screen door, running his fingers along the length of it. Then he dropped down to a squat and lifted the doormat. There it was.

As soon as he stepped inside, feeling his head and shoulders move past the doorjamb into the different light and different smell of the inside of the house, he felt it again, something that was both anger and pleasure skittering around on the skin of his shoulders and his upper arms. It always made him think of the first time he'd felt it.

He and Lewis had seen these Arabs leave from the apartment where they lived behind their grocery store. They always left a window high up in the side wall open.

But why was he so preoccupied with the notion of the first time—the first time an animal was killed, the first time *he* killed an animal, the first time he'd gone into a house? The first time was what made the difference. It looked forward, and after that all the other times looked back and remembered it. When you went into something, you didn't know what it was going to be like, and the first time filled you up with what it was like. After that it was just like putting a little more air in the same balloon. It made a difference in what the rest of the world looked like, too. Things looked different after the first time you did something, but after the other times you knew how they were going to be.

The window high up in the side wall was open the way it always was, and they had seen the whole family leave, piling endlessly into their car, all of them talking at once, the fat woman he took to be the mother pushing and pulling the kids, yelling at them in between talking to the others. Now they were gone and the dark grocery was empty, and so was the apartment behind it, even though they had left some lights on and a radio playing. People are fucking dumb.

With one boost from Lewis he was holding on to the windowsill, and a second or two later he was crouching inside the frame of the window, looking down into the dark inside of the store for a way to climb down, waiting for his eyes to get used to the dark, to pick out the precise spot to put his foot that would take his weight and yet not knock anything off the shelf. Climbing down the shelves silently was awkward but not hard, staring at boxes of cornstarch and bluing (what did people use bluing for?) while his toe reached for and then found the next shelf in the darkness below him, but when he got both his feet on the floor he realized he'd been holding his breath. All his muscles felt good and tight as he started to move along the aisle, his head and shoulders leaning forward slightly into the next step, into the next second, till he glimpsed in the half-lighted volume of air above the shelves and a little in front of him somebody moving furtively.

For an instant he stopped breathing, wanting to run forward and backward at the same time, wanting to fight and deny everything at the same time, before he realized that it was himself he had seen moving in the big round anti-shoplifting mirror, himself moving among these shadows. That was when he first realized that he'd been feeling it all along, the feeling of *difference*, ever since he'd gotten his body completely inside the window. The mirror only showed him the half-dark inside of the store from an angle that was almost exactly the same as when he had crouched there inside the window frame waiting for his eyes to get used to the dark. It was an old empty cube of a room whose interior volume was charged with darkness like a velvety dust and with silence like a pressure in his ears. It was the same store he'd been coming to for years to buy chocolate milk and those small packaged berry turnovers, and to try to steal packs of gum or a candy bar from the racks in front of the counter while the Arabs were distracted by one of the other guys back by the Coke cooler, but at the moment, crouching above the top shelf and looking down into that dark geometry of shelves and counters, plate-glass windows ribbed with Venetian blinds, some dusty light filtering through them from the streetlamp outside, boxes and bottles ranged along the shelves, he could not have imagined a space as strange as that. And there, barely defined in the strangeness, the foreignness of that dark aisle, he saw himself looking up into the mirror with his mouth open, being there in a place that stubbornly continued to be empty in spite of his being in it. These were the dark aisles that the Arabs walked through, casually, at night, after closing, if they needed something, a can of corn, a stick of margarine, frozen pizza. They walked around in here then without even thinking about it.

He moved quickly to the counter and slid behind it. They had not even bothered to lock the cash register. He pressed the NO SALE button while leaning his stomach against the drawer, hugging the machine with his other hand, trying to smother the ringing of its bell. He took all the bills, making a small pile of them in his left hand and neatly folding them in half and sliding the wad deep into his pants pocket. From the shelf beside the register he took a half-pint bottle of Southern Comfort and put it in his back pocket. Then he moved to the back, to the door that led to the people's apartment.

And that was different too. The fact that the lights were on and the radio was playing made it seem even more strange, like another world or time, some other life. Other people lived their life here, something that absolutely excluded him. They had a calendar on the wall that he could not read; in the sink and on the drainboard next to it were dishes with the remains of stuff in them that he could not recognize as food. He knew he was wasting time, that they didn't keep valuables in the kitchen, but he almost couldn't help himself, it felt so good on his skin and around his head, being in this place where people lived, where they took for granted all these things that were so strange to him, so foreign—the color of this table, the way these chairs were spaced around it, the radio that was playing now that was addressing itself to them, to the Arabs, the way a radio playing in another apartment or even the next room doesn't sound like it's talking to you, except that now he was *in* the other apartment and it still didn't sound like it was talking to him.

As far as this house was concerned he didn't even exist, he thought as he unplugged the TV set, winding its cord around the handle and putting it next to the front door. In a corner of the living room a small table was covered with papers and envelopes, business stuff, and he went through them quickly, without really expecting to find anything. There wasn't any money, but he took a small box of blank checks.

Going down the hall (he was moving faster now), he just pushed open doors and poked his head into the rooms. He didn't bother with the kids' rooms, but in one of the others he found a drawer with piles of what he took to be jewelry. He did not know how to carry this stuff away; it would never fit in the pockets of his blue jeans. He yanked a pillow off the bed and pulled off the pillowcase. He got the earrings and necklaces, and some cufflinks and a ring out of another drawer. He also took a small radio out of one of the rooms in back (somehow he felt that he should not take the one that was playing), and a camera. He didn't find any more money, but in the dark bottom of one of the drawers, underneath some clothes, he scooped up a bunch of little foil-wrapped packages that his hand recognized as rubbers. He heard himself snort a kind of *Humph!* as he dropped them into the pillowcase even as he was already moving back toward the hall.

He could smell his own sweat now like a zone around and surrounding him as he moved through the rooms of that apartment, through the various smells that belonged to this place, so different from his own. He realized that he had never once stopped moving—except for the moment when he'd seen himself in the mirror—the whole time he'd been in here. The back of his throat was dry and he was having trouble swallowing, but he knew he was moving back toward the kitchen now by the sound of the radio that had told him all the while he'd been here which direction he'd been moving in, even though it wasn't "talking" to him, even though there was nobody home.

Now as he passed the kitchen he wanted to stop and get a drink of water. Or even better, a beer. To get a beer out of the fridge. To open the door of the fridge casually, the way the people who lived here did, without any haste at all, not even really "hearing" the loud *clunk* of the door handle as his hand pulled it down and held it open while he bent down, finally dropping to a squat to check if there was any beer. But he knew he must not. A glass, a beer bottle, the door handle of the fridge: they would take his fingerprints.

He was also unaccountably tired, his arms and legs feeling heavy, his throat and chest constricted, so that he was having trouble breathing: the pressure of the darkness he'd felt when he first looked down into the grocery store from the windowsill was now a pressure that filled the apartment itself, making it hard to breathe. He had no idea how long he'd been in here. It seemed so long that he could barely imagine a time when he hadn't been here or when he would be outside again, not hearing this radio.

Back in the living room he stopped for a moment and

forced himself to stand up straight (had he been stooping *all* this time?), forced himself to stand still. He could control his legs, even though they felt as if they would begin trembling any moment now. What was it that he wanted to do now? To sit down on this couch and smoke a cigarette. He realized he regretted now that he'd unplugged the TV, because he actually wanted to sit here and smoke a cigarette while watching something on TV and flicking his ashes into this ashtray with incomprehensible writing all over it. The way the people who lived here did. But he could not force himself to do it. He really didn't have any idea how long he'd been here or when they would come back. At any minute he might hear their car pull up outside.

He fumbled for a cigarette, right where he was standing, and put it in his mouth. But he could not hold his hand still long enough to get it lit. Finally he waved the match out and dropped it into the ashtray. Then in a fit of exasperation he picked up the ashtray, hefted it for a moment, then dumped the butts and ashes onto the couch and stuffed it into the pouch in the front of his sweat shirt. He could feel it against the muscles of his stomach as he leaned down to pick up the TV and the pillowcase, and he could hear the people asking themselves with a kind of outraged incomprehension, "But why would he take an ashtray? It's not even worth anything!" That would mystify them and infuriate them even more than the money or the TV: the absence of the ashtray would insist on his invisible presence in the room, in the house. Like a ghost, he would haunt the place where they had their lives.

It was two days later (they had not been able to resist talking about it) that Jerome took him aside and told him, "You are a fool."

"I am not a fool."

"Yes. Both you and Lewis are fools, but you are the biggest fool."

"How do you figure?"

"About six or seven different kinds of fool."

"How do you figure?"

"I figure it like this: you went inside and he stayed outside and looked out, right? Right there you are both fools. If you were caught, you would be charged with burglary and he would get off as an accessory, if he even waited around to get arrested with you. I really don't think he's *that* much of a fool. But look at how much of a fool he *is*: he still doesn't know how much you really got out of there. He has to trust you that you have given him a complete accounting, and I really don't think that *you* are that much of a fool—to take ninety-nine percent of the risk and give him fifty percent of the profit."

Albert looked at him.

"Humph," Jerome said. "I didn't think you *were* that dumb. But maybe you have a fool's courage, too.

"Look," he went on. "You are taking a big risk any time you are breaking and entering, so you have to make that risk worthwhile. A day's gross in a little grocery store like that is—what? One hundred, one hundred fifty dollars, tops. Do you really want to do time for seventy-five dollars?"

"We got a TV set and jewelry, too."

"Shit. You will be lucky to get fifteen, maybe twenty-five dollars for the TV, if you can find a safe place to sell it. And I'll bet you money that the jewelry is all dime store. And where do you get this *we? You* went in and got it, even though you had to split the take with Lewis. See, you have not done any time in prison, but I can tell you, you do not want to take that kind of a risk for this amount of profit. It does not make sense.

"But there is more foolishness in this thing yet. For instance, you did not even look around behind the counter for the man's revolver, did you? If you had gotten the piece, you would not have been able to keep from bragging about it. You did not even look around for the strongbox where they'd be sure to keep two or three days', maybe a week's worth of gross. That would have been in the house part, probably in a desk drawer or someplace close to a desk. You did not look around for a suitcase or satchel to carry the small stuff out in. No. You grabbed a pillowcase. If *you* saw someone walking down the street carrying a TV and a loaded pillowcase, what would you fucking do? But if you were carrying a suitcase— You see what I am saying?

"Did you think of an escape plan? If those people had come home while you were there, how would you have gotten out? Sure as hell not through the door they are standing in. Sure as hell not out through the front door of the grocery, which the old man has about seventeen locks on. You have to do some advance planning if you are going to do this seriously—and if you get caught you will find out plenty fast how seriously you *should* have been doing it. How much good do you think Lewis would have done you—standing around outside with his finger up his ass—if those people had come home while you were in their house? What was he supposed to do, whistle?

"Last of all, you were probably walking around all the time you were in there on probably the only valuable thing those people have. You were probably walking around right on top of five or ten times more money than you took out of there."

He paused and Albert looked at him awhile before he asked, "They keep money under the floor?"

"No, fool, the *rugs*. They're Arabs. They get rugs direct from Arabia, from their folks in the old country. Don't you know what those rugs are worth?"

"Humph," Albert said. "Who is going to buy a rug from me?"

"I'll show you," Jerome said. "You may be a fool yet, but maybe you have a fool's courage. I will show you how to make the most of it."

Jerome was smart. He had been in prison three or four times. The first thing he said was "Forget all this cat-burglar bullshit." After dark was the worst time to get into houses. People were much more likely to be home or come home, and neighbors more likely to notice lights being turned on or a flashlight being used in a house where people were supposed to be out. Much better to go in during the daytime, when people were at work and you could see what you were doing.

Everything had to be predicated on avoiding even being seen. But the odds all went against that, so you had to play the odds. Given that you will be spotted eventually, how do you avoid actually being *taken*? All you can do is be ready for it.

"Like a fucking Boy Scout, Albert, you have to *be prepared.* See, if somebody comes home and suddenly finds you standing there in his living room unplugging his stereo, he's not expecting that. He's not ready for it. He is going to be shocked. But if you are prepared you can take advantage of that shock and get out of there even before he can recover enough to get a good enough look at you to be able to describe you to the police. 'What did he look like, sir, tall? Short? Brown hair? Blond? Mustache?' 'I don't know. It all happened so fast.'

"First thing you do is wear a hat. Any kind of a hat, a watch cap or a baseball cap, a beret—anything that will hide your hair and maybe even part of your face. Then you want to wear a jacket. The best kind is those reversible ones, different colors inside and out. Then under that you wear some kind of a shirt that's a different color from the jacket. A nothing-colored pair of pants and your basketball shoes: you may have to break some track records." Albert looked at him.

The rest of it all came out slowly over the days and weeks and months that followed, as they were doing the houses, sometimes four or five a day, and then they would lay off for a while. Always, before they went into a house, Jerome would talk to him, telling him what to expect, what to look for, the escape routes and the meet-up points. He would explain it all slowly, then go over it again as they drove past the house. Albert would nod yes. Afterward Jerome would talk to him again, explaining the reasons behind the changes, if any had been necessary, pointing out mistakes, but also telling him when he had done good. They would sit in Jerome's kitchen with their feet up on the new TVs or stereos, drinking ice-cold beer out of the bottles, Jerome doing most of the talking, dropping his ashes on the linoleum. "What I like about this job is the hours."

One of those times Albert had asked, after drinking a couple of beers, "What would you do if someone came through that window right now"—All he could see through the window were the tops of the buildings across the street and, several blocks away, a billboard showing a blonde woman in a black velvet evening gown drinking a glass of whiskey. She had a broad, Swedish-looking face and wore her hair short—"and tried to take this stuff?"

"I would walk calmly over to that drawer over there," Jerome said, "take out my thirty-eight-caliber revolver and tell him to halt. I would aim it right at his chest and watch him wet his pants. Then I would waste the fucker's ass. Nobody steals my goods."

Albert watched him pointing his gun finger at the person in the corner. He watched him flip back the thumb of the "gun" twice, firing silently, then dropping the pointing hand slowly, following the dead person as his lifeless body crumpled to the floor. Albert looked at the color TV he had his feet on. On the screen a sexy woman was talking silently about Tampax. She was wearing a tennis outfit and sitting in front of the mirror of a dressing table in a bedroom bigger than any he had ever actually seen. He looked at the corner, where the dead body would have been. He knew Jerome would do it; he had already seen him kill that little black dog, bashing out the life in its body with a brick. The body of the person lying there in the corner would also have been full of life, would have had a life of some kind. What kind? A mother who nagged him, an itch to get laid. Would also have talked about houses he'd done: "See this gold chain," lifting it with two fingers up off his collarbone.

"The people who owned this TV," he told Jerome, tapping the set lightly with the back of his heel, "could have done that to you."

"Or you," Jerome said, laughing and flipping his ashes on the floor.

Jerome's .38 had come out of an apartment Albert remembered as particularly messy—mattresses and TV on the floor, cheap stereo on an orange crate, empty six-pack cartons littering the place, wastebaskets spilling their old Kleenexes and cigarette butts and matted lumps of hair, ashtrays overflowing, drawers gaping open with various pieces of clothing hanging from them. On a sort of chipped-up nightstand next to the mattress in one of the bedrooms, still in its holster, was the revolver. It smelled of oil and was loaded.

"Next to the bed," Jerome had said. "This sucker would have used this on us." He picked it up and handed it to Albert, saying, "Here, you take it. I'll take the next one."

But Albert had turned it down, and it ended up in that drawer. Jerome had taken it along a couple of times, which made Albert nervous, and when he finally left it in the drawer Albert felt better. He tried to imagine shooting somebody, an actual person. He tried to imagine the man Jerome had described standing there in the corner, backing away from him, his hands held out in front of him as if to ward off a bullet, as if they *could* ward off a bullet, backing away and pleading, pleading to be let to live, and then that look of panic coming into his eyes when he realized that this would be, for him, what all his life he had called "dying," a word that up till that moment had had no meaning. It was then that the man wet his pants, peeing on himself the way an animal would that also knew its life was going to end. The man was still trying to protect himself, trying to say "No" to death—not making any sense—and when he saw the hammer start to pull back on the gun he started to turn toward the wall, looking over his shoulder at the gun, which went *pop*, and sent him jolting into the wall as if somebody had kicked him in the side once, one good one, and just as he was about to make a face as if to say "That *hurt*," "something" kicked him again in the back, just under the shoulder blade, and he ended up lying there with one leg bent underneath him and his face against the wall. It had been hard enough killing the animals, and Albert was glad that Jerome didn't insist on bringing the gun with them anymore.

He had almost not been able to stand it when Jerome had killed the big black talking bird. As soon as it spotted them, it started yelling its head off, yelling, "Pirates! Mr. Bill! Pirates! Let me out of here!" and whistling these piercing shrieks. First Jerome had grabbed a big towel and covered its cage. But the bird kept on screaming—stuff you couldn't even understand now, but still one hell of a lot of noise. That was when Jerome started to look like the "ice" man. He went over and ripped the towel off the cage, which only made the bird yell louder. He took the cage off its stand—it was huge—and carried it back into the kitchen. The bird had a yellow beak the size of a hunting knife and Albert knew that no way was Jerome going to stick his hand into that cage. He took it out to the back porch where the people had a big white freezer chest. The bird was just about breaking its wings against the cage and peeing and shitting all over itself. When Jerome closed the lid, you could still hear the bird yelling, but muffled. It almost made Albert sick to hear that animal that wanted only to live, that was now slamming its body against the bars in there in that freezing darkness, though by the time they left the house it was silent.

Doing houses with Jerome, he made fairly good money, and Jerome taught him a lot as they went along. He learned about tools—that the police would consider even a flashlight and a screwdriver "tools" if they caught you carrying them together. You could carry just the screwdriver or a tire iron or some other kind of small crowbar by themselves, though. And you did not even need the flashlight. He learned how to check out both the house and the neighborhood. Ideally, you would have two escape routes. He learned how to get the names of people off the mailbox. Then you looked up their number and called them to make sure no one was home. If somebody answered, you just asked for Seymour or some other dumb name, and got told you had the wrong number. Sometimes, Jerome told him, you could actually go back and do the same house again. You waited two or three months, till the people had done all the insurance stuff to get the TV and stereo replaced, and then you hit them again and got *brand-new* stuff. "You see what I'm saying? You do a repeat performance when they least expect it—dig? When they least expect it."

They rode around in Jerome's car—or the car Jerome was driving; it was a different one every few weeks, always with out-of-state plates—looking over houses and neighborhoods, Jerome doing most of the talking, telling him that he had to think all the time about how to maximize the profit and minimize the risk. Once you had broken and entered, you were in for a felony. If you had to bail out in a hurry because the people came home, you had better make sure you had made it worth your while. So the first rule was: "Get something small and valuable and get it into the car *immediately*." Then even if you had to run at least you had gotten *something*. Then you went back in and tried to complete the job. But you always wore a watch, and you always kept track of what time you entered, and you made certain you got your ass out of there in under thirty minutes.

In those first few months they had only two close calls, and since he and Jerome really were prepared they were able to get away clean both times. One of the rules was: "Drop the shit and run. No color TV is worth two years." He was busy in a corner of the living room of this place, pulling the wires out of the tape deck, when he heard a car pull up outside. Jerome was actually at the front door, one hand on the knob and a box of silver under his arm. He put the box down abruptly on the floor and said, "People. Split." They both walked quickly out the back way and kept on going, through the backyard and over the fence to the next street. As they walked back around the corner, Jerome began telling him about a movie he'd seen, interrupting the narration at times to tell him—in precisely the same tone as the movie conversation—"People who live on this street, in this neighborhood, they take it for granted. They don't look around while they walk along. They *know* this street. So you want to look like you know the neighborhood as well as they do. Like you live here. So you get yourself all involved in a conversation and you *can't* look suspicious."

Albert had been listening, and only now realized that

Jerome had led him back to the car, parked right in front of the house they had just been in. Jerome quietly pressed down the trunk lid as he slid past it and got in behind the wheel. As he started the car, he was saying, ". . . and then the leading lady, who has tits out to here—" He put his hands on the wheel and pulled slowly out of the parking spot. Albert could not resist looking back, looking at the house. Through the front window he saw the woman, her back toward him, standing, looking at the remaining components of the stereo where he had left them lying on the floor. She had her hands on her hips. He could not see the man.

"All you have to do," Jerome was saying, "is keep your cool. There isn't anyone else's cool you've got to keep, and there isn't anybody else who will keep yours."

The other time he was actually putting a TV into the trunk when the man drove up. Albert got into the car and started the engine, but waited till the man had pulled all the way into the driveway. Then he honked the signal and pulled away. When he got around the block to the meet-up point, there was Jerome in his jogging suit, "jogging." He got in the car, saying, "Just keep driving slowly. Head for the freeway." As he was talking, he was pulling off the shiny jacket.

The trouble with doing houses with Jerome was just exactly that efficiency that made it actually profitable. They were in and out of the houses so quickly it was only in bits and pieces that Albert could taste again something of the feeling he had experienced in the Arabs' store and apartment. Now and then there would be a moment, walking into a bedroom and realizing it was a child's bedroom and yet standing there for a long moment, longer than he needed to, noticing a big white plastic goose standing on a shelf, and then noticing the cord that connected the goose to an outlet, and pressing the switch on the cord and being pleased to see the goose light up. A lamp in the shape of a goose. It pleased him. He thought about the child who lived its life in this room. A boy, he decided. He thought about the people going to the store and seeing that goose lamp and deciding that their child would like it. They must have talked it over. But then there was Jerome.

Another time there was a weird thing for matches—made out of pottery—with wooden matches poking out of the top. It was sitting on a coffee table. Around the base of it was printed, in big red block letters, CASSIS QUENOT, and below that, in smaller black letters, something about Paris. These people must have brought this thing back from Paris, France. When they wanted to light a cigarette, they reached over to this thing they had brought back from Paris—with what casualness they reached out to it now—and pulled out a match. He put it in the pocket of his windbreaker. He would not bother showing it to Jerome.

Then there was the time they did the condo. He had a funny feeling about the place as they walked up the sidewalk (stepping around a pile of wet dogshit). And sure enough, as they were pulling and hauling on this huge TV—they were right in the middle of the back doorway—this guy came in the front door. The guy yelled, "Hey!" his mouth going very wide just then, and began waving a very big, heavy-looking gun. Albert turned and ran but Jerome had to jump over the TV set to get out the door, and Albert heard him grunt as his feet hit the concrete and took off running, and then he heard the shots. He made it around the corner and was just grabbing the door handle of the car when he looked up to see Jerome step in the dogshit and go sprawling. His feet went out from under him and he came down hard on his tailbone. Albert was behind the wheel now, wrenching the key over in the starter. It looked like Jerome had had the wind knocked out of him: he was trying to get up and not making it past a sitting position. The man came around the corner now, still carrying that huge gun, just as Jerome got himself over on one hand and his knees. Albert heard the shot and saw Jerome's body give a jolt at exactly the same moment, and he stepped on the gas. He thought he heard another shot after that, but he couldn't be sure over the noise of the engine. As the car lurched into high speed, he felt a tingling in the skin of the back of his neck and scalp, his shoulders, and a vague sick sensation in his abdomen, like diarrhea.

He never saw Jerome after that, and he laid off for more than a month. But then he saw a house that was perfect—it had a hedge that hid one of the front windows from the street—and of course he had to do it. He went over all the rules he had learned and went ahead and did the house. It was the bottom half of a clapboard duplex whose inside still smelled of fresh paint. Albert's eyebrows went up and he felt his face spread out in a smile when he saw they had their stereo components still in their boxes.

In the kitchen he remembered something Jerome had told him and he looked in the cookie jar and in the set of cannisters that said FLOUR, COFFEE, TEA. He reached his hand into a box of oatmeal and pulled out a heavy silver necklace with big blue stones all over it. He carried it in his hand into the bedroom and started to go through the drawers. As he was scooping more jewelry into a shaving kit, he caught a glimpse of himself in the mirror. He saw himself standing in the bedroom where these people slept and fucked. They had just recently moved in. Tonight they would come home to find they had been ripped off. The police would come and take all the information. After the police left, the man and woman would be sad, but they would comfort each other and then they would get into bed and make love. There was the bed. They still had some framed pictures sitting propped up against the wall in various corners.

He picked up a small toy seal made out of some sort of tinny metal. The seal had little wheels, and when you rolled him along on top of the chest the ball on the end of his nose turned. He put the seal in his pocket and went back into the kitchen, where he took a bottle of beer out of the fridge and tried to open it, but his hands were shaking so badly he had to try three times, and even then he cut his knuckle. He pressed on the cut with the dish towel he'd been using to open the fridge and hold the bottle. His knees were shaking and the muscles in his calves felt light and fluttery, but he leaned against the counter and picked up the bottle and lifted it to his mouth and drank out of it.

It was not that he was any more afraid of being caught than at any other time. He wasn't afraid at all. It was how intensely he felt the thrill of being in the house. It was a place to live, and partly it was so delicious because it was so absolutely forbidden, much more forbidden than the stealing, this *being* where other people have their life. He felt their cold beer fill his mouth and slide down his throat.

He thought of the framed pictures leaning against the walls. They had brought them here from where they had lived before, insisting that the house become part of their life, the wall they hung their pictures on. And yet all the walls were already here, the rooms laid out and named—living room, dining room, bathroom—before they had ever gotten here. The more he thought about it the more excited he became, having to lean again against the counter and take a long swallow of beer. They were trying to make this house become part of them, and the whole time it was standing all around them, containing them inside itself, holding them in its own body.

He was stopped two days later by a policeman, who wanted to talk to him about the oversize stereo speakers in his back seat, and arrested. That was the first time he'd been made to lean all his weight on his hands spread out on the fender of his car, with his legs extended and the cop behind him kicking the insides of his ankles to spread his feet farther apart on the pavement. He had been pulled over into a bus stop, and some people were gathered around watching as they waited, and when the bus came they got on.

He went to prison several times, but he always went back to doing houses. He made good money at it and he would tell people what he liked about his job was the hours. They allowed him to do the things he really wanted to do—hang out with his friends, play pinball and basketball, get laid, go to movies in the afternoon, which he loved to do: to come out of the cool darkness of the movie, still excited with the other life he had been part of, and then to walk out into the hot, bright light of the day. He felt then that the day was something he had chosen, as he had chosen the movie.

It was while he was playing basketball one day that he had gotten a taste—just the palest shadow or suggestion—of that feeling of foreignness, of difference he felt when he was alone in someone else's house. These guys from some other neighborhood showed up to play, and when they all started in to choose sides, one of them said, "It doesn't really matter: if the score is too lopsided, we will just trade some personnel around."

Albert looked at him with his mouth open, then closed it but could feel his face beginning to frown with puzzlement. He asked finally, "If you are on the winning team, why would you want to trade away your players?" He was almost angry.

In the houses the foreignness or difference was actually a heavy charge in the air of each room. When he left, he always felt light and strong. And the difference felt stronger as he penetrated deeper into the house that stood all around him silently, or was inhabited by the voice of the radio. He could feel it even in the living room or the kitchen, where the people ate their breakfast without thinking about it very much. When they were someplace else, not here, they could think about this kitchen table, but their thought of it did not include him being here in this place, while his thought of it did include them.

But those rooms often felt like the outside of the house, the skin or some layer of tissue just inside the skin. Sometimes he would do nothing for a while but stand in a hallway or on the stairs, a place without windows, some spot that gave him the feeling he was deep in the interior of the house, where he would look—first without really seeing—at whatever they had there: a photograph of a big church, documents, family pictures. The people in these pictures were part of the past of the people who lived here. They had a life that went back into a time when people wore strange clothes and drove cars that looked like antiques. They had children too, whose lives moved forward into a time he could not see, like the time that passed in the world while he was in prison, a time that excluded him, for which he did not exist. Then he realized that he was looking at a single panel of a Dick Tracy cartoon these people had framed and hung here, and he wondered if it might be valuable, some sort of collector's item, so he took it.

On the top shelf of the closet in one of the rooms he found a box containing a small plastic Baggie of marijuana and a long pipe made, evidently, out of ceramic, in the shape of a penis. Whoever lived here put the marijuana in the bowl of the pipe, a shallow well back near the balls with a sort of sieve-like screen at the bottom of it, and sucked the smoke out of a hole in the head end of the penis. Somebody had actually thought of this thing, and then had actually made it. That person knew while he was making it that someone would buy it, that it would go from his factory someplace to a store, where it would sit on a shelf till someone saw it and decided to buy it. It would have been the girlfriend of the woman who lived here. She had gone into one of those porno bookstores with her boyfriend. She had suggested it, laughingly, but with an element of a challenge in the suggestion, and they had gone in, where she laughed too much and too loud, and then she had seen the penis pipe and said it would be perfect for this woman. They had all had a good laugh when the woman who lived here opened the package and saw the penis. They had smoked a little of the grass in the pipe, the boyfriend at first saying he would pass on this one, until the girlfriend teased him, saying, "Yeah, you men. You want us to suck on yours all the time. If you think the real ones are so hot for us to suck on, why can't you even put your mouth on a fake one? It's just a *pipe*. Or are you afraid we will think you are latent?" Finally the man, to be a good sport, gave in and took a couple of small puffs, awkwardly, from the head of the pipe.

When her friends had left, the woman had started to put the pipe away, thinking, Where can I hide this thing? What if someone should find it! But as she stood in the middle of the room with the box in her hand she could not resist opening it to look at it one more time—it was so *funny*. Then she took it out of the tissue paper in the box and hefted the weight of it in her hand for a moment. She was still pretty high. She put the box down on the dresser and got ready for bed, feeling a vague excitement diffuse itself throughout her body. As she sat on the edge of her bed rolling up her hair, she could see in the mirror of the dresser how raising her arms like that lifted her breasts. They would look more attractive to a man like that, the nipples seen by him through the sheer cloth as fully extended and pointing up. Of course, that would be why pinup girls were posed like that, with their arms raised. Something like that, some detail like that, which seemed so small, the fact of her breasts being lifted, the fact of their being seen like that, moving slightly inside the cloth that hid them and revealed them at the same time—that would be enough to make a man excited as he watched her, his penis inside his shorts beginning to grow stiff. When she had finished with her hair, she went over to the dresser to put the pipe away, but instead brought it into bed with her, where she began by putting her mouth on the head end of it, the way she had done to get a puff of grass. The glaze of the ceramic was cold. Then she mouthed as much of it as she could, getting her saliva all over it, and finally inserted it into her vagina and brought herself off. In the morning she put it on the top shelf of the closet.

Holding the box, Albert felt as real and as solid as the dresser or the desk that had watched the blanket over her pelvis pluck up each time her hand came back and up underneath it to pull that thing out of her before she stroked it in again until her knees came up, making a tent of the blankets that hid what she was doing, and at the same time he felt like a ghost, present in the room yet nowhere to be seen. Even as he pictured what went on in the room, he did not see himself in the scene watching. He had become only a consciousness in the room or a consciousness of the room, part of the body of the house.

There were other times, too, when he felt as if he would disappear just as he stood motionless in a hallway or on the landing of the stairs, feeling the house around him that somehow denied him, whose life excluded him so thoroughly. He would see a place where the paint had chipped off a corner of the bannister and he could see three or four different colors of paint. The house, he realized, had gone through a whole series of lives, different families coming in and painting it different colors, and the whole house remembered and preserved those lives, and all those lives excluded him. He would feel a panic then, an anxiety whose source he could never name except that it was connected with the house, with his being in the house, as if—unless he did something—he would dissolve, simply not be there anymore, like the smoke from a cigarette, or that he would faint. It scared him to feel things he could not understand, like now this heavy feeling of loss. What was it he was grieving for?

At times like this he would swear to himself never to do houses anymore, not if it made him feel like this. Then one day he was in a house and he felt like going to the bathroom, deciding suddenly to find the bathroom and pee in the toilet that belonged to these people, feeling then the old excitement—using in this way this place that formed the frame of an entire life for these people. But on the wall in front of him as he peed there was a note someone had tacked up, and the note made him feel that other thing, the anxiety and loss. Then he felt a cold object in his hand which held his attention through a sort of personality, and when he looked down he saw it was the handle of the toilet as he was about to flush it. He felt as if he wanted to throw up, it was so hard and cold. Then as he moved over to the sink he saw himself in the mirror over it. It was him. There.

After that he always found the mirrors first, just in case he needed them, and the houses stopped scaring him. He even discovered he was doing things he never would have thought he could manage. Like the time he went up to the second floor of this house and heard noises coming from above the ceiling. They were not human noises—a hum, and with it some other noises, like some sort of machine, repeating themselves at regular intervals. He forced himself to follow them. They were coming down a narrow stair-well that probably led up to an attic. You could not tell what people would put in an attic they had converted—TVs, stereos, tape recorders, movie projectors. Looking up the stairs, he saw the lights were on. The noises were something small, like a sewing machine, and then over that steady humming there were these other, regular clunks and bumps. He could feel his body beaded with sweat all over underneath his clothes as he deliberately moved his feet up the steps without making any noises. He knew he was breaking one of the basic rules—you never got yourself into a room that had only one exit—but as he climbed the stairs, more and more of the attic came into view above the stairwell: the joists of the ceiling and the studs of the walls lit by small hanging lamps that threw weird shadows. Whatever was making the noise was coming from all around the attic, not a sewing machine but something spread out, something that took up the whole area, something that must have been on top of the tables he now saw set up on sawhorses, the tables set together with their edges flush so that they made a sort of second floor in the attic, about waist level, and in the dimness underneath that level he saw a forest of those sawhorses. There was no one in the room, which was totally taken up by an enormous model-train set—miles of tracks that went by billboards and forests, mountains with tunnels through them. Some of the tracks went in between the beams that held up the ceiling, the roof of the house, and through towns set up with depots and even a post office; *Coming Home* was playing at the local movie theater; a school yard was laid out with a couple of basketball courts, even nets on the hoops. A whole world up there in the attic, one of the trains still running around the outermost track, solitary over there on the far side of the room. In all the streets and the depot, in all the stores and all the switching stations and water towers, even in the logging camp in the mountains, there were no people. The train was running now behind some low hills, invisible, and he alone heard it there as it moved.

Sometimes the houses would have signs—usually on the gate to the backyard—saying BEWARE OF DOG. After he discovered that some of these places didn't even have any dog, he systematically checked them out. He almost always avoided the ones that had animals, but the worst of it was that some of the places that actually did have them did not say anything about them. It occurred to him that some of the people bought the dogs to protect their homes and property, but others put up the signs without the dogs, trying to protect their property with words.

Once he went into a house he thought was empty: he had seen the widow lady who lived there go off with her shopping cart. He got in the back door, which she had left open, and was doing fine (TV out of the living room, silver out of the dining room) and was headed down the hall when suddenly a large man's back appeared in front of him. The man had lurched out of a doorway Albert had not even known was there and turned back toward the kitchen—actually turning his back on Albert. But it was too much of a startle, the man just suddenly being there, and stumbling, too, doing a sort of Frankenstein-monster walk, stiff-legged, with his arms out in front of him, losing his balance and caroming into the wall. Albert heard himself yell and drop the box of silver. The man stopped and grunted, "Mama!" and then lurched around, a bright smile on his idiot's face, a gross, misshapen face, round and hairless and pink, drool spilling out of the corner of his mouth. Albert felt himself backtracking, amazed to find himself thinking *at the same time*—This is one of those Mongolians! This thing would start in to yelling, and how long the widow woman had kept this thing hidden in here. Her secret, huh? The secret life of the house that mouthed now silently, *Mama*, and then sat down—it must have been forty years old: its fucking hair was salt-and-pepper gray—and actually cried, blubbering. Albert picked up the box of silver, opened it and put it in the thing's lap. The man stopped crying now and began pawing awkwardly at the bright pieces.

In closets and drawers he found the secrets of other houses, too—a stack of porno postcards, a whole drawer full of dildos and vibrators, a life-size woman in a closet who scared him into a silent scream in the half-second before he realized she was only inflated rubber. In a shoe box on the floor under one woman's bed he found magazines filled with pictures of young men jacking off. He found drug paraphernalia, and in a house that belonged to a judge he found hundreds of little potted plants that he recognized as marijuana. Then there were wigs and false teeth, a closet full of broken artificial legs, tubes of Preparation H in the drawers of nightstands, bottles of hair dye and whole boxes of tubes, compacts and bottles of stuff he called makeup. Sometimes in the back of a cupboard or the top shelf of a closet he would find packets of paper with writing on them, writing that obviously had never been seen by anybody but the person who had done it, lines of writing that did not even reach the right-hand edge of the page, like a grocery list. Others would be bundles of letters, sometimes tied with a ribbon, sometimes with a rubber band. Secret writing.

He found that he had become something of an expert on the cost and value of the various brands of stereo components. In a house that had probably the most expensive stereo setup he thought he had ever seen, he saw also the most elaborate fish-tank outfit—more than anything he could have imagined. There must have been thirty different tanks arranged on shelves that came up the stairs, all lit from behind, all the tanks complicated by pumps and bubble gadgets and thermometers. He looked at the fish open-mouthed. Some of them were transparent, some of them streaked with colors he had never seen, others veiled in fins that draped around as they swam like some sort of ballet costume. They were not like the other animals. He did not think of them as animals. They moved in a way that was totally different, in a medium that was totally different. As he passed them again and again on his trips back and forth to the car, they irritated him more and more. When he had gotten everything there was to get (even a coin collection out of a back closet), he went back and looked at the fish. Coldly beautiful and delicate, requiring all this amount of care. For what? Bullshit. He pulled the lid off one of the tanks, then turned on a small radio on a table just inside the living room and threw the radio into the water, listening to the sparks and fizzle and watching the fish jerk around a little before they went limp and floated to the top. The lights and bubble gadgets in most of the other tanks blinked off. He went down the line of the tanks still functioning: where he could figure out the temperature controls he turned them up to boiling, and where he could not he just yanked the tubes and wires out of the water. One enormous tank he couldn't do anything with and saved for last. He picked up a pottery elephant with a sort of castle on its back from a corner of the living room and threw it at the tank, forcing himself to stand there long after the crash, watching the cascade of the water, and the fish with it, as it poured whitely down the steps to the front door.

More and more now he found that after he had gotten his one valuable thing out of a house and into his car he wanted to go back in the house and simply stand there, feeling the house as a kind of ongoing zone around him. He had been taking things that had no value from the very beginning, even though he did not know what he would do with them. Outside the house they had come from, they lost whatever magic they had seemed to possess—a spoon that said NEW JERSEY, a small plastic syrup bottle in the shape of a bear, a half-size railroad spike that someone had had brass-plated. They were things that had acquired a sort of magnetic charge from having been in one place, where people actually had lived, for a long time. Outside of those places they seemed not inert but diminished. Some of the older pieces he eventually threw away because they had lost whatever it was that had prompted him to take them in the first place. He had forgotten which houses they had come out of. But he always took the money things—stereos, TVs, jewelry. They formed the public reason for his being in the houses at all. How else could he explain, even to himself, that all he really wanted to do was *be* there in a place where people had their lives, and where they *had* had them, so that the place was somehow also a gathering of time? How could he explain—if he was caught—what he was doing feeling real in someone else's house? The things—and even the smells—that were gathered here spoke of things done by these people who referred easily to the things the others had done, and also to each other's plans, things that were going to be done. These references folded around the place a feeling he could never describe in words but that he knew he liked, and needed to feel.

One day, passing an APARTMENT FOR RENT sign, he was struck by an idea that seemed at first totally crazy, but that would just not go away. He wanted to break into the vacant apartment, just to see what it was like, just to stand there, to be in it, the way he was in the other houses while he was robbing them. But would this be different? And how? He was actually more nervous getting in than he had been any time in recent months. (Finally he decided that if it came to that he would tell people he was interested in renting the apartment. The thought of that made him give a snort of laughter.)

It was one of those railroad flats—a long hall moving straight back from the front door in one unbroken shot. Standing at the front door and looking back through the apartment was like standing in a square tube or conduit that was ready, waiting, to convey some sort of fluid. But the place was empty. Doors opened in the right-hand wall onto rooms. In the front was a room where a fireplace had been converted to a gas heater. Above the mantel was a mirror in which he saw himself standing in a vacant apartment. Behind him was the blank yellow of the wall, and two squares of different yellow hovering there as the ghosts of two pictures. The sliding doors to the next room stood open. One cubicle came off the hallway just far enough to enclose a toilet. How strange to have a special little room to do nothing in but pee and shit. Next door to the toilet was a regular little bathroom, except that it had no toilet. It made sense, he thought, but he had stood in all these rooms, every one of them, slowly turning his head and even turning his body this way and that, and it was not the same.

People had lived here, and they would live here again, but the fact that nobody lived here now spoiled it. He thought that what was so exciting about being in the other houses where people were living was just the fact of standing in the middle of someone else's life. Like the time he had actually sat down in the kitchen and made himself some toast. It had been all he could do to swallow the bread, and it had damn near messed him up good because some neighbor woman had seen him through the window and called the police. Up until he saw the woman peering at him through her window with the phone in her hand, he had managed to control his anxiety about feeling like a ghost by glancing from time to time at his reflection in the chrome of the toaster, inside of which now the heating elements were toasting this day's bread for him. This thing had been given to these people as a wedding present by the wife's father, who was hopelessly hooked on gadgets. It could toast four slices at a time and had no handle to pull down: you just dropped in the bread and a spring sensed the weight of the slice and lowered it quietly into the guts of the machine, where it was toasted. The bread had been bought the day before by the husband, who had made a special trip to the store because he was the one who had forgotten to put it on the shopping list. Their daily, ongoing life. He sat in their kitchen, sweating and tense and feeling good feeling himself being there. In the chrome toaster he saw a flattened-out fun-house version of himself bring the cigarette to his lips and take a drag. He picked up a knife with a paper napkin and buttered his toast, making sure the butter went all the way to the edges of the slice. He wanted to stay, somehow to be here while these people were doing this, just this, just to be here, to become the table or the toaster and in that way to participate in what went on here. He put the cigarette butt out in the remains of the cube of butter as it sat on the table in the general litter of plates and cups. At first the butter-with-the-cigarette-butt-sticking-out-of-it was simply lost in the overall welter that would be seen as the breakfast mess. But gradually it emerged more and more as the focal center, a kind of understated outrage around which the rest of the details ranged themselves.

But here in the vacant apartment there *was* no ongoing life to stand in the middle of. If there were memories of the people who had lived here, the memories had gone into the walls and floors somehow, were embodied in the choices of paints and wallpaper. The house itself could not remember anything, and he started to leave. But just as he did he realized that this view he was now getting of the apartment was exactly the view each of its tenants had had as they moved out, and also the view they had had of it when they first moved in. It was the bare place, without any accommodations, that was both the introduction and the farewell to the life lived here. The blank, uncompromising thereness of the place enclosed all the other memories of the place as it had been lived in. And this was what he was getting now, the maze-like pattern of empty boxes (which was all the rooms were, he told himself), empty boxes connected to one another, that had been chosen as the place where these lives would be lived. And again he almost had to laugh at the realization that every tenant who had ever lived here, and every tenant who would live here in the time to come—they would all have this same view of the blank apartment stretching out in front of them as it now stretched out in front of him. Except that he saw them as they looked down this hall or walked quietly into the rooms, their shoes echoing in the hollow shell of the house. But their view of it did not include him, him standing here, being here, stubbornly, looking the place over as if he would actually decide to rent it. Even in the midafternoon the place was dark. The interior of the apartment had only two windows, and each of them looked out into an airshaft. What fool would want to live here, he thought.

What would it take to make this place livable? How many of the things that weren't valuable that he had taken from the other houses would he have to bring in here, and how long would he have to leave them here before the place acquired the feeling the others had? He did not know why the question made him angry.

He was still angry a minute or so later, when he decided to leave, at first starting to move back toward the window of the service porch that he had forced open to get in, then changing his mind and walking straight down the length of that hall to the front door, which he wrenched open. He found himself looking directly into the face of a young man in a suit and tie who was holding a key aimed right at the place where the lock would have been. Behind the man was a middle-aged couple.

"Hello," the man said. "Did Mr. Jurgensen send you to see about the stove?"

Albert looked at him, enraged. In a minute he would push this fool out of his way. "What?" he said. He could hear the open contempt and anger in his own voice.

Now the man was confused. "I thought you were the stove man," he said. "Isn't that your truck outside?" Behind him the couple was busy looking politely interested.

As he stood there in the doorway, Albert could feel the vacant space of the hall as it stretched itself the length of the apartment and exerted a pressure on him, on his back, even though he could not tell if it was pulling him back into its elemental blankness or pushing him out, excluding him again. The man in front of him was probably close to six feet tall, a little shorter than himself, but he looked lighter. Albert looked him in the eye and said, "No. I am not the *stove man.*"

"I don't understand," the man muttered. Turning to the couple behind him, he started to "explain": "I thought—" and then turning back to Albert he said brightly, "Oh, you must be here to look at the apartment too. Why don't we—"

"No," Albert said. "I am not here to look at the apartment, I live here. I own this place."

The man in front of him could not any longer pretend he did not notice the hostility. "Look here," he said, taking a step back and suddenly becoming very "official," "I'm going to have to ask you what you're doing here."

"I am here to tell you to fuck yourself," Albert said,

taking a step forward and standing too close to the man. Albert could see the man's face register the realization that if he did not give ground he would have to fight. The man could not keep his cool, Albert thought, simply because he had crowded him, simply because he had taken away some of the space the man had assumed was his to stand in. If the man had only kept his cool when he stepped back, he might have been able to salvage the situation, but now he was just confused, and afraid of getting hurt. He did not even know how much he might get hurt or if there was any way he could still avoid it. He was angry at being shamed in front of customers and the adrenalin was flowing in him for sure, but that only made things worse for him. Albert recognized the look in his eye from the animals. It was hateful to see, painful to see, and he turned and walked down the porch steps into the sunlight, leaving the three of them standing there next to the open door of the apartment he had just been inhabiting.

He was still keyed up when he drove past a house he remembered as a place he had broken into before. When had it been? He pulled around the corner and parked. Yes, it was just about three months ago that he had been parked across the street from this place, setting up to do a house on the other side of the block, when the people started coming out of this brown shingle place, yelling back and forth that they would meet back here at nine o'clock. Of course, then they left the lights on and the radio playing. Smart. Now he drove around the corner and backed into a parking spot almost exactly in front of the house. Next to the brown shingle house was an old clapboard, and next to that was a pedestrian sidewalk that cut through the middle of the block to the next street over.

The place was perfect. Looking through the window, he could see the brand-new stereo, right where he remembered the old one had been. To get to the backyard, you walked along a little alley that ran beside the house, where the garbage cans were. From the front you could hardly see the gate at the far end of this walk because of the overgrown hydrangea bush. He pushed on the gate but it would not open. He reached up to the top of the gate and climbed over. The sucker had put a big lock on it. He had also put some kind of little lock on the window of the breakfast room that was screened from the backyard. As Albert pried open the window with his tire iron, he watched the screws of the little lock pull out of the wood, and wondered how much the poor fool had spent on it.

When he got to the new stereo, he remembered what it was about this place that had pissed him off. These people were too damn neat. The man had installed the components, and then he had kept the wires running together in neat bundles tied with plastic fasteners that looked as if they had been made just for that. The wires were then tacked to the walls with those white insulated staples, the bundles of wires following the lines of the shelves or the corner where two walls met. Like fucking ants. His tire iron would not fit under the staples to pry them loose—and this sucker with his new locks on the gate and on the windows and his tight-ass wire installation, this sucker was just costing him *time*. He swore softly to himself and got a knife from the kitchen. He cut all the fucking wires and lifted out the stereo. But when he got to the front door with it, he saw that the man had also put one of those double dead-bolt locks on it. The back door too. He had to go out the window with the stereo. That did it: he would clean this bastard out. He would not leave him a bar of soap. Then there was the new lock on the back gate. He put down the stereo and pulled the tire iron out of his pocket. Now the bastard's two cats were nosing around the stereo and looking up at him nervously. He waved them away with the tire iron and turned to the gate. It was a very impressive-looking lock, but the turkey had attached it to a very old redwood fence with one-inch screws. They pried out of the wood like loose teeth.

Back in the house he paused for a minute trying to figure out the most efficient way to strip the place. He began to pile everything just outside the breakfast-room window. When he was upstairs getting the TV and a smaller radio, he glanced at the bed and remembered the man's set of cassettes. He would take them too. He grabbed a pillowcase, but then had to laugh at himself. He rummaged through the closets until he found a suitcase. He swept the cassettes into it with his forearm, wanting to make up for all the time he had lost, and then quietly remembered: "All you've got to do is keep your cool." He got up with deliberate slowness, closed the suitcase and put it by the back window.

He went to the fridge and opened it with a dish towel. He looked at his watch. He got out a beer, and as he leaned back against the fridge drinking it, he saw for the first time a door that went off the laundry room. They had a whole other room back there, a room he had not even known existed. He opened the door and found the light switch. God damn!

He was in a room that looked like a small recording studio—white soundproofing stuff on all the walls and ceiling, carpet on the floor and all over the place panels of switches and gauges, tape machines of all sizes and descriptions, things with TV screens—some of them small, and some with round green screens—lots of microphones and cable and even a big thing with a piano keyboard in the middle and a switch panel on the top, like a telephone switchboard.

This was some kind of jackpot. But what the hell did they do in here? They must be musicians: on a white table in a corner he found lots of music paper covered with indecipherable notations that didn't even look like notes. He stood in the middle of the room and took a long pull from the beer can. This room was some kind of secret. You could hardly tell it was here from the outside because it didn't have any windows. It was hidden away in the back part of the house, back behind the laundry room, and yet it was obviously an important room. It had the same feeling as a repair garage. It was a place where work got done. But it was also like a hospital, like an operating room. The work that got done in here was highly technical, and *clean* in a way that irked him, even though he could not say why. There was also a smell in the air that he associated with electronics, with computers and transistors and high voltages running around behind gray metal panels. It was separate from the rest of the house—cut off from the house and sealed off from the world too. He realized suddenly that in here he would not be able to hear if anyone came home, and quickly went back to the main part of the house. He decided he would pack up the car with what he had already stacked up outside the window, and then if he had any room left in the car he would start in on the stuff in the secret room.

The idea of a secret room started his heart pounding again. What did these people do in there? Maybe they were spies. Bullshit. They could be making bootleg records. But he had not seen any records in there. Bootleg tapes. That was it. The man had some dumb-ass job in an office somewhere, but when he got off work he would go back to that secret room and become a bootlegger. It was secret in a different way from the jack-off pictures and the drug paraphernalia. This was something that stayed a secret for a while and then emerged from that unseen room into the public world, where it circulated, freely and openly, even though its true, undercover nature, its stolenness, remained concealed. It made him giddy to think he had been standing, standing in the middle of that life—his own, invisible secrecy wrapped up this way in someone else's. He went into the bathroom and looked at himself drinking from a can of beer in this bootlegger's house. He imagined phone calls the bootlegger would make to Detroit or Atlanta, and now the thrill of being here in the house merged into a different anxiety. If these people came home and found him here, they would not call the police. They might shoot him quickly or—

He poured the rest of the beer down the sink, threw the can in the garbage under it and put the dish towel on the drainboard. He climbed out the window and picked up the tape deck and the typewriter. At the far end of the little alleyway beside the house he paused for a moment, hidden by the hydrangea bush. A man was coming casually down the sidewalk. He turned in at the steps of this house. Fucking jackpot. Albert laid the goods down on the ground and walked out from behind the bush. The man was—what, about forty?—built O.K. He looked pink, his face just a little flushed, so wrapped up in his own life, not knowing he was being watched. Coming home to his own house where he lived his life was something he did day after day—*day after day!* Albert suddenly wanted to kill him, to beat the life out of that smug body, to destroy the body, to make it all go away so that there would not be any of it left, not even a memory.

Instead he focused his mind on the escape plan. He was walking straight toward the man, who now looked up as he climbed his own front steps. A look of puzzlement came over his face for just a moment. He stopped, looking Albert in the eye, and asked, in a voice whose sarcasm made Albert furious, "Can I *help* you?"

"No," Albert told him, continuing to move now toward the neighbor's lawn. "You cannot help me. I was just back there."

The man had continued to come closer. He was not afraid. "Well, what the hell were you doing back there?" He was demanding an answer. Now would have been the time to have the gun. Now would have been the time to push the big barrel of the gun into that face whose pinkness came from all this, this house, this "property" whose boundary he was sure he was now stepping over, going onto the neighbor's lawn. He would have pulled back the hammer of the gun, saying, "I was back there playing with your toys, with your illegal bootlegging stuff, because *this is your life*. How did you like it, chump?"

As he stepped onto the other lawn, he was turning his back on the man, saying, "I was just back there talking with your neighbors." Now he was walking across the neighbor's lawn and in about three seconds he would be around the corner of this house in the pedestrian walkway. The fool would think he had run up that walk to the next street, and he would go inside his house. Then he would freak out and call the police. But Albert would be waiting just there, and as soon as the man went inside his house he would stroll down coolly and get into the car.

He stood by the mailbox, listening to his heart going. He exhaled hard, and then stood silent for a long time with his head bowed before starting to inhale—slowly. He forced himself to do this four or five times, thinking that it was all those fucking locks that had slowed him down. He ought to go in there and take those locks off that sucker's doors and shove them up his ass. He heard a woman's voice behind him ask, "May I help you, young man?" She was standing behind the half-open front door, with just her head sticking out.

"Everybody wants to help me," he said to nobody in

particular as he started to move toward the car. As he got within ten feet of it, he saw, out of the corner of his eye, the pink-faced man coming down his front steps—*toward him.* He thought of running for it, but figured he could probably beat the man to the car, and that was the best bet for getting away. He did not mind fighting the man and was pretty sure he was not heeled. As he got into the car and began to start it, he saw in the mirror the man behind him, the man responsible for all this. He thought of putting it in reverse and running him down. Suddenly the frame of the rearview was totally filled with something white. What the hell was he doing back there? The trunk lid, of course. The son of a bitch had opened the trunk and was stealing back his stuff. Albert rammed it into drive and floored it, peeling away from the man in a squeal of rubber and a blare of horns as two passing cars swerved to avoid him, the trunk lid bouncing open crazily behind him.

2 / KINDERTOTENLIEDER

When he got home, his face felt a little hot and kind of tingly, so he looked in the rearview mirror, and sure enough, he *had* gotten more sun than he thought. His muscles felt all loose, pleasantly tired. A white Datsun down the street had done a terrible job of parking, its nose end poking diagonally out into traffic. He had done his eighty laps and felt as if he had finally really gotten himself back into shape. When he looked up the steps, he saw a tall young man in a shiny warm-up suit coming out from behind the hydrangea bush that screened off the alleyway leading back to the garbage cans and the gate to the backyard. This guy is obviously casing the place, he thought, but what the— "Can I *help* you?" he asked, almost snarling.

"No," the man said, "you cannot help me. I was just back there."

"Well, what the hell were you doing back there?" David demanded. The man was walking away from him now, across the Sanders' lawn, toward the cut-through. He was obviously trying to get away. But what the hell can I accuse him of, David thought.

"I was just back there talking with Jim Nabors."

Jim Nabors? David thought. He wanted to push past the

hydrangeas, to get a look at the back gate. If that was open— But then he almost stumbled over his own typewriter and tape deck.

Suddenly he was *very* excited, almost possessed with haste. He ran to his front door but he could not sort out the keys in his pocket. When he did find the right key, he could not get it into the lock. He had put these damn locks on the doors and now he couldn't get into his own house. He was frenzied with the idea of getting inside to the phone, to call the police, but his hands were shaking so badly he couldn't get the key into the lock, and meanwhile this guy was probably halfway down the next block. He looked over toward the Sanders' lawn, in the direction the man had gone, and was startled to actually see the man again, walking down the steps of the cut-through, back toward the sidewalk, toward the white Datsun he had earlier thought so terribly parked. David went out onto the porch to see the license plate, but for some reason he could not read it, even though he was looking directly at it.

Now he became obsessed with getting that license number. He walked briskly down his steps, watching the other man also walking toward the car. When David got to the rear bumper, he saw why he had not been able to read the number: the man had draped a washcloth over the plate. A corner of the cloth was still poking in under the trunk lid, as if it had somehow been *accidentally* pulled out of the trunk, and had somehow fallen into this position. He had to admire the man's imagination and attention to detail even as he yanked away the cloth, only then realizing that the trunk lid had not in fact been latched. It stood ajar now, and through the opening he could see part of his stereo. He yelled out, "What!" and pulled open the lid. He reached in and got his turntable, all the time reminding himself to keep his cool, to stay cool enough to remember the number, and repeating it aloud to himself over and over. Just then the car pulled away down the street, the trunk lid bouncing up and down violently. And as the car receded he stood there on the sidewalk holding his turntable out in front of himself, wanting to say something to the man, hearing himself yell out, "Z-N-two-four-three-seven!"

Later he would tell this story, and he would laugh at himself at this point. "I don't know what the hell I meant," he would say. "Maybe I wanted to tell him, 'I've got your number!' But I didn't want to take any chance of forgetting it. Actually, I ended up really liking that repetition: I used it in this new piece Anselm and I are supposed to have finished."

"Is that the *Champs Magnétiques*?"

"Right."

"When are we going to hear that, anyway?"

"Well, you don't just *hear* it. You've got to get inside it. It's an *environment*. You've got to get inside it and walk around."

"Well, when are we going to get to walk around inside it?"

"Oh, you know—"

The worst part about getting robbed this time was—"Well," he would say, "eventually every part seems like the worst part. The cost is really nothing, because of the insurance. But what if Danny had been there? Or Jane? What if the guy had been armed? The feeling of being violated, of having your privacy invaded, that's pretty bad."

"I came in," Jane would add, "and found my underwear strewn all around the bedroom. He had been into *every-*

thing. And then when the cops came *they* walked through, and here are all my bras and panties lying out in the open. It was like being violated twice. You feel you have no secrets left."

"And then we had to tell the summer tenants, the people who were going to take our place for the summer. The guy hit us a week before we were scheduled to go back East, and these total strangers now have to be told they're going to be spending the summer in some sort of high-crime area. But it worked out: we didn't tell them about the first burglary—which I think was done by the same guy. He came in through the same window, he went for the same stuff. But anyway, that's why we ended up installing the burglar alarm."

The house Anselm had gotten them for the summer had also worked out—a big rambling old two-story brick house painted white, with what seemed like a whole collection of roofs, all of them metal, all seeming to slope at different angles. The only problem with it was the bees. The house stood by itself at the very edge of the woods, at the end of a gravel road, and somehow the bees would get into the kitchen, and then they'd beat their brains out against the windowpane trying to get out. Every morning somebody had to sweep up a couple of dozen dead bees.

But it was great for Danny, who had the whole woods to explore, the whole world of nature right outside the door, all the way to the bank of the Hudson itself, only two miles away. And in the other direction, about half an hour by foot, was the campus of the college, with six newly resurfaced tennis courts and a student union where Danny immediately figured out how to beat the two old pinball machines, so that even though they were no challenge at all he was able to show the other kids—mostly summer kids his age in for the music camp—how to beat them, and so made some friends, with whom he wandered around the woods setting off firecrackers. Sometimes he would bring some of the kids home and jam with them if they played the right instruments and were into rock, but most of them were super-casual friends who changed every two weeks, with the sessions of the music camp, and most of them were pretty straight classical types. On any given day you would run into two or three of them practicing in the woods—not just flutes and clarinets but trumpets and trombones, and one day David ran across a skinny girl with frizzy hair playing her cello right in the middle of the path. She had carried her little folding canvas chair and her music stand, and was sitting there serenading the beeches with something largo from the Brahms First. At least she was playing it largo. It would have been wonderfully romantic if she had been any good. But still it was pleasant. Danny made a great show of being scornful of these kids, saying, "If we had a long enough extension cord I would bring my axe out here and blow these turkeys away." But the next day as David was walking down the gravel road—just to clear his head—he heard the sound of a blues harp coming out of the woods, and when he got to the clearing there was Danny, sitting with his back against a tree, practicing a three-bar figure over and over again. "Hey, Dad," he said, standing up and putting the Hohner in his back pocket, "want to play some basketball?"

François was the only local boy Danny's age. His father was a Canadian who had come down from the Maritimes years ago to teach physics at the college. They knew Anselm, but only casually. They lived in an amazing rabbit warren of a house right on the bank of the large creek (the Sawkill) that flowed into the Hudson. The father was endlessly in the process of building it, using only the logs and stones of the area, but cut and joined and finished with a remarkable craftsmanship. All the exposed wood and stone and all the indoor plants made being in the house barely distinguishable from being outside it.

Jane said, "It's a lovely house to visit, but I don't think I could live there. I need a little more of a feeling of civilization, I think."

"Amen," David said.

"And the sound of the waterfall twenty-four hours a day would drive me bananas."

"Does François play an instrument?" David asked Danny.

"Yeah. Tennis racket."

"Is he any good?"

"He's not bad. He's going to teach me how to hit an overhead."

François also had a good collection of comic books. The only time Danny really felt lonely was in August, when François and his family went back up to Nova Scotia for two weeks to visit relatives. That was when Jane saw the archery set at a garage sale, and figured it would be good for Danny, and for them too. It was something they could do together because it was new to all of them and they would all be starting to learn it at the same level, she said.

It was François who solved the mystery of the secret room for them, but that was much later, just before they went back to California. David had noticed one day as he drove up to the house that from the outside it was bigger than all the rooms accounted for. There was some kind of superstructure tucked away at the back of the building, with its own metal roof that sloped at yet another angle from the others, but that structure somehow disappeared when you were inside the house. It was the kind of thing he told himself he would look into one of these days, but never got around to, telling himself perfunctorily that it was an attic or a storage room of some kind, and letting it go at that, except for those times when he noticed the odd superstructure as he drove up to the house or when he came back to the house by the path that went past the refuse dump in the woods on the way to the river. Then he would tell himself that he wanted to check it out. But almost always, as soon as he got inside the house, he forgot about it. He was much too preoccupied with the piece he was working on with Anselm.

About every three or four days, he would drive over to Anselm's studio, or Anselm would come over, and they would have what David called "show-and-tell." If he went over there, Anselm would start out turning everything on at once until David could calm him down to one projector or one set of monitors at a time. "This sequence," Anselm was saying, "doesn't work unless you've got six monitors. I want to *force* the viewer to make those comparisons: 'Are they in phase?' 'Did that sequence repeat this one down here *precisely?*' See what I mean?"

"Yeah. I like it. I like it. I think."

On one screen a field of wheat grew, in a matter of a minute or so, from single fine spears of vulnerable green seedlings like chives to a sea of green waving tufted plants, and then ripened to amber, still waving and rippling as the wind moved over it. Then the sequence started over.

"See, right there," Anselm said, pointing at the screen as the wheat began to change color and ripen, "right there I want to dissolve in a sequence of actual waves or sea surface, right? This viewer is going to have to work his ass off. No rest for the wicked. And while that's happening, see, this one over here is tracking on the cemetery—see how those rows of tombstones *flow* over that hill?

"But what I really want you to see," he went on, leaving

David standing in front of the panel of monitors, "is this little beauty. This mother took me three weeks just to do the wiring—"

"I don't like the cemetery," David said, without turning around.

"This circuit random-cuts between six different cameras, and of course you can set it up for six monitors or six optical printers or six projectors—any way you want to go. So like—here," switching on another video monitor, "you get six different takes on the same image. Which of course makes it six different images, right?"

On the screen an amazingly beautiful young woman's face was looking directly at him, first saying, "Kiss," and then pursing her mouth in a kiss. The cameras had been arranged all around her head and as she spoke and kissed, David saw her from a random selection of the six camera angles. "Kiss." The screen was filled with the blonde back of her head. "Kiss." A three-quarter angle from the left. "Kiss."

"Who the hell is *that?*" David asked.

"Why don't you like the cemetery?"

He loved the mornings, those heavy wet summer mornings of the Hudson Valley. Often he would wake up very early, in the blue-gray light of dawn, lying for a long moment on his side, gazing out the bedroom windows that looked back up the long straight white gravel road flanked by the woods on one side, conscious of Jane beside him, her body heavy with sleep. He would lie there listening to the birds and, under that layer of sound, the occasional groan or creak of the house, his wife's breathing, an insect humming. He would get up then, quietly, and take his clothes downstairs. The house was so delicious then, when everyone was asleep, leaving it to him, and he loved being able to leave it, to go out onto the porch naked and put on his clothes out there, feeling the soft airs of summer on the skin of his body. Then he would walk, either through the woods to the river or up the gravel road toward the college. The world then was being given him, in the creak of the branches as he walked past them, the soughing of the leaves, the crunch of the gravel under his Keds. He walked all the way to the college, realizing as he drew nearer that he actually *was* hearing angelic voices. It was the kids from the music camp. They were in the chapel practicing Fauré's Requiem. He wanted to get closer, to hear it better, but they were so far away, the voices so faint, that every step he took on the gravel crunched out the music. He ended up walking a few paces and then standing still to listen. He wished he had thought to bring the tape machine. He would bring it tomorrow. He would track in on the music, finding the exact place on the road where he would first be able to hear it, the precise threshold, and then track in, getting closer but leaving the volume steady, so that the listener would feel himself moving toward the source. It would be hot tomorrow; the shoulder strap of the Uher would cut into his shoulder, but it would feel good to be getting to work this early, while the rest of the world was still asleep.

When he brought the tape machine the next morning, it bothered him at first that the microphone wasn't directional enough, and was picking up the crunch of his feet on the gravel as well as the music. He could have fixed that with a shield or baffle of some sort, if he had thought about it. But that *was* part of the world he was recording. He had so often to remind himself that the incidental noises, the background or ambient noise level that we ordinarily tuned out as well as we could—that was part of the experience of hearing anything. Like Cage hearing the random traffic noises as part of the string quartet he was listening to.

He walked steadily forward, holding the mike out in front of him, his head, adorned with earphones, bowed to watch the VU meters, thinking also that he must look quite weird, like a witch doctor or something, walking down the road at this hour, holding this strange forked wand out in front of him. He knew also that when he heard the playback of this tape he would be able to see this whole morning: the white road, the soft light, the moving wall of trees on his left, the open field on the other side of the road, the blue sky of a summer morning. But what would the listener see? That was Anselm's job, to give the listener something to look at. As the voices swelled a little, he thought maybe a squeeze shot of some leaves, the camera pushed up to them so tight that the screen was filled only with a green geometry of veins and edges of leaves, then pulling back to reveal, eventually, the whole of the woods. "The forest and the trees," he thought. As the viewer's eyes were telling him he was getting out of the woods, getting a more and more encompassing, more commanding prospect, his ears would be taking him deeper and deeper into the field of the music, the domain of the music, the beautiful voices singing about death. But that was not really his concern. Anselm would take care of that.

He saw the VU meters start to climb and heard the car engine in the headphones at the same time, and looked up. Oh damn. He did not want to go all the way back to the threshold point he had found back there and start tracking in all over again. Could he just switch off and start it all over again from here? He could include the car noise in the tape. It *was* part of the real world he had been recording. He could also just edit out the noise later.

He sat down on the porch swing, laid the cassette machine on the seat next to him and adjusted the headphones carefully over his ears. Then he turned on the machine. Danny was setting an arrow on the grip of the bow and pulling back the string. He looked tall and lithe, built like a basketball player. Jane stood beside him and a little bit behind. She had stuck a feather into the back of her hair, and when she saw David had noticed it, she clapped her hand over her mouth a few times, smiling. They were standing on the front edge of the lawn, fifty feet or so from the target, which they had set up directly against a tree, at the line that marked the beginning of the woods.

"You're going to lose your arrows!" he yelled at them.

They turned toward him and opened their mouths, saying something he could not hear. Danny's arrow moved in what looked like slow motion, leaping out and up and then settling down again into a sort of groove, and entered the target somewhere close to the bull's-eye.

He loved to work on the porch, which felt like being in the house and outside it at the same time. He had done some of his composing here, but it was too distracting, finally, and now he saved that for the studio down in the basement (*cellar*, he reminded himself), where it was cooler and almost totally soundproofed. But often he would dub a mixed tape that he wasn't quite satisfied with onto a cassette, and then take the Uher up onto the porch, sit on the swing with his headphones on and run it through.

Now he was in that non-space the headphones created. His left ear was listening to waves crashing, with some inland birds calling during the quiet phases and a very high soprano sax playing a staccato lead line laid over the whole thing. His right ear was gripped by a synthesizer generating a repeated syncopated three-bar figure over and over again. The bass was too heavy on this channel and he had changed his mind completely about those crashing waves. There was no way to fix them, no way to make them be felt as ironic. They had to be cut, but that would leave a hole in the texture at this point. He could not think right then what he would put in that place.

Danny walked up to the target, gave him the thumbs-up sign, which he returned, and pulled the arrows out. They had all been tightly bunched around the center, even though he had not hit any bull's-eyes. Behind him David could feel the house, the solidity of its bricks and its roofs, the big chunky square spaces that its rooms coolly enclosed. It was a great house for the summer. It dated back to the eighteenth century, which probably explained some of its quirks—the way all the rooms on the ground floor were rather small with very high ceilings. The kitchen had actually been separate from the rest of the house, though some later owner had built over and around the breezeway, joining the two parts with a short hall, in the same style and brick as the rest. Then the kitchen had been redone recently: the brick had been exposed (probably re-exposed—in the other rooms the brick had been plastered over), ugly cabinets built in and all the usual convenience and status gadgets installed.

The spaces of these rooms somehow enclosed or included history, too. People had probably sat on this porch wearing silk knee hose and white wigs. Maybe it had been built by descendants of the original Dutch settlers, some burgher who made a handsome living from the sawmill he had built on the stream where François's father was still building his house. In fact, that house *had* been a sawmill or the site of one. The people would eat in the dining room behind him wearing their frock coats and their lacy cuffs, the ladies' brocaded gowns sparkling in the candlelight as three young boys played a Mozart trio, the cadences mixing with the clink of the silver on the china and the conversation.

"Tea?"

"Yes. Tea."

"In the harbor. I don't believe it."

"But it's true, I tell you."

"But it's mad. Dressed up as Indians!"

Mozart. Mozart might work too. Foss had done it with Bach in the *Baroque Variations.* Bach's notes ended up sounding like ocean waves, white breakers of sound crashing on some shore. But he wanted to keep the definition of the notes. He could hear the Köchel 282. The kid would have written it just about the time these people behind him were having their conversation. He would have been what—eighteen? He pulled the headphones off. Now he could hear the K. 282 on tape coming over the big speakers in the studio downstairs, the little room gripped in the clarity of that lyricism. He would stop the reels and hold the tape between his thumb and index finger and pull the tape quickly past the playback head, watching the real-time indicator go crazy, hearing all that lyricism flatten out as the notes staccatoed their way out of the machine with no interval at all between them in a hammering, metallic technoscream. He would have to transpose it all down about two octaves, but it might work. Anselm would make his usual noises about "stealing," and he would have to explain about stealing. He turned off the cassette machine and stood up. Danny called out to him, "Come on, Dad, see if you can do any better!"

David waved him away and lurched into the house. He sat at the piano picking out what he could remember of the opening of the sonata. It was in E flat. He tried it down two octaves, and then sat with his hands in his lap trying to hear *that* as it would all come crowding out of the speaker at three times this speed, with the notes stacked up on top of each other the way things looked in a telephoto shot. It sounded a little like some of Nancarrow's things for player piano, but of course the Mozart had a melodic basis, a solidity of understructure.

He looked around the room, only then realizing the magnitude of the change that had come over him when he walked through the door back into the house, something that skittered around on the skin of his shoulders and upper arms. Outside it was hot, bright and sticky; in here it was cool, dim and comfortable. The crisp whiteness of the walls, the classical proportions and the accents in delft blue, the open lid of the baby grand—they all spoke of civilization, of the straightness of lines and the geometrical arrangements that civilization imposed. Outside his wife and son were shooting arrows and facing a solid wall of woods. That wasn't any wilderness, by a long shot, but it sure was the opposite of this. Even calling the trees a "wall" was an attempt to impose on them a kind of order that had never resided in them.

But then if the house imposed an order on the world around it, it imposed that order also on the people who lived here. In this living room or parlor, which was on the small side anyway, they had tried to crowd a sofa and a coffee table, a set of speakers and, along with the baby grand, a set of drums and a chord organ. Lost in a corner was Danny's amp. Being in the room was like trying to relax in an elevator. What the hell must these people have been like?

Anselm knew them, he didn't know how well. The father had taught in the history department here. History of Law. The wife was a tennis player. She had also done things like organize a women's chorus for the faculty wives. They were all, obviously, a pretty musical bunch. The son had been something of a prodigy—played all kinds of instruments, wrote poetry, was some sort of whiz kid at electronics, too. He had actually put together the studio downstairs. David was certain the money came from the wife: you sure as hell were not going to put together a studio like that on a history professor's salary. The boy had been about Danny's age—twelve—when he'd been killed in an accident. A year ago? Two years ago? Then the father had gone completely to pieces and eventually got involved in a shooting. David had never gotten all the details, and it was of course the details that would bring them to life. Now they were just "the Mortimers," a name that preserved only a shadow of their actuality in a kind of limbo in which they were neither forgotten nor realized.

He went into the kitchen and got a beer out of their fridge. Were they the ones who had chosen these cupboards? Ugh! They couldn't have been all that filthy-rich, because someone had done a fairly amateurish job of installing the dishwasher: the side panel did not come flush with the wall. A couple of bees were buzzing into the windowpane. He leaned against the counter and picked up the bottle and lifted it to his mouth and drank out of it. He felt the cold beer fill his mouth and slide down his throat. He was curious, standing here in their kitchen, what their conversation must have been like—at breakfast, say. How would you talk to a child prodigy? (Danny was a bright kid, and a pretty talented musician, but—) It must be like dealing with another species, like talking to a Martian. You really could not get more foreign, more *other*.

But then what must it have been like for the boy? Mozart had been pretty freaky, but only in one area. This kid had composed *and* written poetry *and* known enough about electronics to put that studio together. It had taken Anselm three weeks to figure out that one random-switching circuit, and this kid, at eleven or twelve, had in effect built a recording studio, complete with mixing capability and oscilloscope monitoring down there under the house. He had even patched together a number of jerry-rigged random sound generators.

Well, where was the stuff? Where were the poems? Where were the kid's tapes? Did he write any music down, even performance notes? Where had they gone? He would have to ask Anselm.

The kitchen window looked out into the woods. Its rectangle was filled with a green fluttering as the leaves, stirred by a breeze, alternately caught and reflected the light that drifted down to them. It might have been a window that looked out onto a light shaft. What did it mean to say a man had "gone completely to pieces"? What did it mean to say the boy had "been killed in an accident"? Just a little way through that dense mass of green outside this window, his son was shooting arrows at a target. What would it mean to be the father of a boy who had "been killed in an accident"?

He put the beer on the table and looked at the bees. Their bodies littered the windowsill and the floor under it. Danny had forgotten the bee patrol again. He thought of Frank, the cat the Riordans had gotten for them. He had had no idea he could get to feel that way about an animal. Daniel and Annie had a friend in the Richmond district with a litter of ginger kittens—"marmalade cats," Jane had called them—and as soon as they saw him they picked out the sharpest-nosed, feistiest little one, and took him home, where he immediately hid in a rosebush. When he poked his head out through the leaves—a marmalade-colored rose with whiskers—Jane wanted to call him Rose.

"You can't call a male cat Rose."

So they ended up calling him Frank, and they still kept a photograph of him in the hall: Frank, lying on a pillow, the afternoon light coming through the windows and through the brass bars of their bed. Even in repose he had some of that sharp-nosed, feisty quickness. David could never have predicted that he would cry the night the McCormacks, the neighbors from behind them, had knocked at ten-thirty. He had been sitting in the living room reading and they had told him, "Your little cat is lying on the sidewalk over there. I don't know if he's dead or anything, but he doesn't look good."

Someone had given him a cardboard box. Frank was lying on his side with his eyes closed, his sharp nose still. He did not have any mark of violence on him, except for some blood that had dripped out of his mouth and thickened a little on the sidewalk under his head, glistening in the hard light of the streetlamp, but he lay there with his different legs crumpled under him or sticking out in a way that nothing that was alive would lie that way. David used the box to pick up the body of Frank, which was already beginning to stiffen, and he heard himself whimpering. He had to tell Danny. A cat. How had he died? It was impossible.

He took the beer and went out onto the porch to get the cassette. He would have to call Anselm to get the Mozart from the music library. He would also probably have to tape it at the college. As he picked up the cassette, Danny called out to him.

Across from the chapel, Danny pointed out a tiny stone building with stained-glass windows that David had simply assumed was some sort of smaller, private chapel. "Have you seen this?" Danny asked.

"Oh sure."

"No, I mean what's underneath it."

They walked over and David read that it was the family vault of the Barstows, the family that had founded the college. They looked in through the barred gate. Danny pointed to a doorway in the side wall, across from the altar. "Know where that goes?" he asked.

"Another vault, I guess."

Danny beckoned him around the corner, where he showed him a hole where the bricks had fallen away, down at ground level. The hole looked down into the cellar of the vault. A short flight of stairs led down from the door Danny had pointed out. The long marble boxes were stacked neatly against the walls.

Danny pointed down into the cellar with a smile and said, "Nineteenth-century recording and de-composing studio."

"Ugh!" David said.

The poems were a gift of love. LOVE SONGS TO DEATH *from her own child. Only for the sweetness of saying "No."*

That was what the poems said when David did find them.

Someone, the Mortimer boy in all probability, had written that across the title page of the typescript of the poems, and had put the black spring binder that contained them on top of a shelf of tapes in the studio down in the cellar, where David found them the same day he totally re-envisioned the piece he and Anselm were working on and decided to tape the voice of the house itself. LOVE SONGS TO DEATH was typed in caps, centered on the page, and everything else was written in what he took to be the kid's handwriting, using a pen with a very wide nib.

He had been lying in bed with Jane in the afternoon. Danny had gone off with François to a place that was actually called Bishbash Falls, and they would have the house to themselves. He had gone down to the studio to work, but knew he would just putter around for a while, till one of them would start it and then they would decide to "take a nap." And sure enough, Jane came down with the mail, calling down the stairs, "Requesting permission to enter the catacombs! Mail call!" The people who had rented their own house for the summer—the Jerk family (that was Danny's name for them)—had written to say that the burglar-alarm people had come to install the system and that it was working fine. And the cats were fine. Oh, and also they had had a small flood in the upstairs bathroom that wasn't their fault and they sincerely believed they should not be charged for it—it had caused only "barely noticeable" water damage in the ceiling of the living room. Bullshit. Edith Sanders had written Jane, and in the middle of her letter gave them another view of the Jerks: that Edward Jerk, the father, hung around on the landing at the top of the stairs trying to get a look into their daughter's bedroom window. What a creep, trying to pry into their private lives like that. Jannie felt violated and of course now refused to do any baby-sitting for them. There was also a good note from Michael Harrison in San Diego. But nothing from the Riordans.

The day was bright and hot, though heavy, and it was supposed to storm up later, maybe tonight. It was pleasant sitting on the porch swing reading their mail with some Haydn drifting out to them from the cool inside of the house like the ice clinking in a glass. Jane was wearing shorts and a blouse he could not really see through but that made him want to try. Her torso was long, and though she complained that her legs were too short they seemed fine to him. She put her hands behind her head and stretched, smiling at him, and there were her breasts. He went over behind her and kissed her behind the ear, saying, "Mmm, you're wearing perfume."

"You finally noticed," she said. "I was wondering what it would take to get you out of that tomb of yours."

"Tomb?" he said. "I'd always thought of it as nice and womby. And speaking of wombs . . ."

They went upstairs and made love very slowly and deliciously in the dimness of the Mortimers' bedroom, with all the windows open but the shutters closed, and afterward he lay there in a half-doze, looking at the changing patterns of light on the ceiling and listening to the sounds of the woods and the sounds of the house. There was one sound in particular, or a group of sounds, he was trying to hear. It was a series of ticks and groans, some of them coming together to make what would have been a perfect fifth, except the high one was a little flat. An imperfect fifth. The house was playing double-stops on itself. His stomach gave a long rumble. He had been hearing that imperfect fifth regularly for some time now, in the morning and again in the evening, but only if he remembered to listen for it. It was obviously the house expanding and contracting with the changes in temperature, and the sound (or sounds) seemed to come from somewhere below him, somewhere deep in the understructure of the house. He told himself he would find it one of these days.

As he lay there on the threshold between being asleep and being awake, the comfortable weight of Jane's neck on the crook of his elbow, he saw something walk across the ceiling. It was not the shadow of anything, because it was brown, as if it were furry, and it walked on all fours, but its legs and its body too were bizarrely elongated. It walked across the green area of the ceiling and then stopped, and then changed direction. It was the image of something, some animal, walking across the ceiling of the Mortimers' bedroom, which he knew was painted white, but which now was distinctly tinged a green, like the color of grass.

It was the lawn, of course. It was the lawn in front of the house, down there below these windows. And that red-and-black thing in that corner of it up there above him was the power mower, right where Danny had left it. The image of the animal walked over and sniffed at the image of the lawnmower. The world of nature is coming over to check out the world of civilization, he thought.

This is a *camera obscura*, he realized. Somehow the shutters, or one of them, were acting as the lens of a pinhole camera. The light of the sun was being reflected off the Mortimers' lawn and their lawnmower and their deck chairs and that groundhog, and converging at the pinhole in the shutters, which then projected it onto the ceiling, which was acting as the focal plane. He and Jane were actually lying here inside a camera, which the house was. If the ceiling had been tilted at the right angle, they would have had a perfect image of the outside world. He felt like Leonardo da Vinci, who had seen people walking upside down on the wall of his room. They had been out in the garden; the light reflected from them had converged at the pinhole in his curtain. No wonder he was excited. "Of course," he heard himself say aloud.

"Hunh?" Jane asked.

He stroked her face lightly and kissed her, and she went back into her doze. He pulled his arm out from under her head.

Of course. *This* is what *Les Champs Magnétiques* was really all about. No wonder he had not been happy with Anselm's visuals. He could not have known what the piece was really about until he'd gotten inside it. He picked up his clothes, wanting to call Anselm, to tell him about this.

But not now. Just then he heard that imperfect fifth again, the double-stop the house was playing, its secret voice. He walked out into the hall, still carrying his clothes, and stood there naked for a long moment, his head cocked, listening for it, those clicks and that groan.

Nothing.

He must have been mistaken. It was too early in the day. The house would not speak again until this evening, when it started to cool off. He looked back into the bedroom. The ceiling was a dull gray, blank. He looked at the shutters. He could not see any gleam of light coming through them, just a general, softly diffused dimness. He went back into the hall, but there were no windows there. He went into the bathroom and looked out the window: that strong, direct sunlight they had had all morning was gone; the day had already started to cloud up. Tonight they would have a storm for sure. Danny had grown up in California and had never seen a full-scale electrical storm. He would see one tonight.

He glanced at the mirror and saw himself standing in the Mortimers' bathroom. He looked at his face. This is what people saw when they looked at him. This is David Lyman, the composer. He made a face. He shrugged his shoulders. You do what you can with what you've got. He heard it again.

He went out into the hall and stood there, listening, feeling the house all around him, somebody else's house that would now speak. Its voice was too far away, it had come from someplace downstairs, but he could not tell yet exactly where. He had been standing perfectly still for a long time when he realized that for some time now he had been looking—without really seeing it—directly at a single panel of a Dick Tracy cartoon that someone had framed and hung there. He could not tell, in this light, if it was an original or if it had just been clipped out of the funny papers. There was nothing in the panel to suggest that the Mortimers had seen it as particularly appropriate to them, just Dick Tracy, in profile, holding a big and very heavy-looking gun, a .45-caliber automatic. Turning only his head, he looked at the other wall, and saw a photograph of a big church. Why? What had prompted these people to hang just these two pictures here, in this hall? The church looked European.

The colored pictures on the ceiling had come and gone. Here on a corner of the bannister was a place where the paint had chipped off and he could see three or four different colors of paint. The house had of course gone through a whole series of lives, different families coming in and painting it different colors, and the house had remembered and preserved their lives.

When he heard it again, he knew it was coming from almost directly below him. He went down the stairs as quickly and quietly as he could, and stood again, still, tilting his head this way and that, just on the threshold between the stairwell and the dining room, glancing from time to time into the TV room and down the stairs that led to the cellar and the studio. The TV room was also a sort of den or library. They had lots of bookshelves, which were neatly stacked with sets of books in identical bindings. They must have been bought at auctions. How else explain the complete works of Twain on the same shelf with Pierre Louÿs and Pearl S. Buck? The father's history stuff took up a small corner, where the desk was, but obviously he must have done most of his work at his office up at the school. David had once looked into one of the man's own books—*Gens and Justice: The Law in Early Roman Civilization*, R. Charles Mortimer, Jr.—and the opening sentences had been all he needed: "It is not music or art or poetry that distinguishes man from the animals, but the law. Man *is* an animal, and only within the structure of the law can he find that protection from his own animal passions that alone allows music and poetry and art to flourish."

But the records were probably the weirdest thing of all. They had a couple of albums of Ives and Carter, but most of the rest of it was dance-party stuff, or "easy listening," the kind of music you would barely realize you were hearing in a waiting room someplace, wallpaper music. There were also sets of "The Greatest Hits of" records, almost all of them big bands—Ellington, Goodman, Glenn Miller. But these were not the original recordings. They were reconstructions by studio musicians. Somebody's brilliant idea. Pay a couple of dozen hacks to transcribe all this stuff from records—including the improvised solos—so that another bunch of hacks, some Ernie Hecksher or John Wolohan, and their orchestras in some studio in Atlantic City could play all those notes right off the page. In the case of Miller it wouldn't make any difference, but why would anyone in his right mind go to all this trouble to turn music into paper? It was like a Xerox copy of the world. And where in the hell were this kid's things? If he had been such an acclaimed composer, who the hell did he listen to? Sure as hell not this José Iturbi stuff. There wasn't even any rock 'n' roll in the house.

Then he realized that this must have been the parents' doing. When the kid had been killed, they must have just about gone underground—to suppress every possible reminder of his having been here. He had become an unperson. His room, when Danny had moved into it, had not even had a picture on the wall. It was just a room with two beds and a chest of drawers in it.

He heard it again, long and deep this time, and unmistakably coming up the stairwell from the studio. He ran down the stairs, trying to control his breathing. This was perfect. He had been concerned that even after he'd located the source of the sound he would have to come down here to get his equipment, and by that time the performance might have been over for the day. He would have had to wait till the morning. But now he put a fresh reel of tape in the big machine and switched on the omnidirectional mike. While he waited, he thought about the crawl spaces under the house, where it might make sense to use them if he had to get in under the floors to place his mike better. But when it came again it was practically saying, "Here I am!" It was coming from a corner of the studio up above the top shelf of the tape-and-book-storage unit, and between that shelf and the ceiling was just enough room for a mike on a stand. He plugged one in, switched the panel over to it and then looked around for something to stand on to reach up there. There wasn't anything, and he looked at the shelves themselves, which were built in.

He climbed them very carefully, pushing some of the tape boxes aside to make room for his hands and feet. It was awkward putting the mike down on a shelf and then using both hands to climb, but he only had to go up a couple of shelves in order to reach the top, and that was when he saw the black spring binder, its corner just sticking out over the edge. He placed the mike and pulled down the book.

"Hmm."

It was covered with dust, and when he opened it and realized what it was, he almost forgot about the sound, the voice of the house that he had finally tracked down and was now about to capture. *Love Songs to Death*, and then all those scrawls. It was like hearing a voice from beyond the tomb. But it was only the fact of his being a kid that allowed him to get away with such grandiosity. Not even John Donne—

Just then the house spoke again, and he leaned over toward the main panel to see what the VU meters were doing. Just fine. This might be all it would say for now, but in the morning he would be ready for it. He would be up before dawn. He still had the book in his hand, and opened it again, just as Anselm poked his head through the door of the studio, saying, "Pretty kinky! Is this how they write music out in California, *au naturel?*"

The next morning he got up before dawn and went down to the studio. He rewound the tape, hit the playback button and listened while he got ready to thread a fresh tape on the other machine. The speakers gave him back the various clunks and squeals of his moving the microphone around. Then there was a long silence. He heard a voice, his own, say, "Hmm."

A sharp breath, his own, blowing the dust off the cover of the black binder. Some rustling of pages.

A long groan, on two notes that weren't quite a perfect fifth, accompanied by a series of clicks or pops, as if the house were cracking its joints. Creaking its joints. A beam or beams of wood that had come from a tree, maybe even from these woods, and that had stood now for two hundred years, unseen, in the dark interior of the house's structure. Its voice.

"Pretty kinky! Is this how they write music out in California, *au naturel?* Look, I came over to bring you a present, but I would hate to break up a tender moment between you and your—uh—equipment."

"I can explain everything!"

Laughter.

"Look what I brought you: a [something mumbled] mike. I think this was developed by the C.I.A. Super-directional—"

"I really want to talk to you—"

"This thing will pick up a whisper from a block away and—"

"I finally figured out why we haven't been satisfied with the piece."

"This sucker is *so* directional—if two people are whispering—a block away!—you can aim it so you'll hear only one of them. Why haven't we been satisfied? See, you aim it, just like a rifle. What's that?"

"The kid's poems, the Mortimer boy, the one who was killed. Hey, what is their story, anyway?"

"Wait a second. What's wrong with the piece? You know, I haven't been entirely happy with it either, to be perfectly frank."

"Oh man, it's been *both* of us. We've both been thinking of the audience as an audience, sitting out there in rows, looking at projections and listening to music that we put on for them up on a stage."

"Do you want to trade places with them?"

"No, but I—"

"It would be a lot less work for us, but I don't know if—"

"We've got to bring them *inside.*"

"If they could handle their end. How do we get them *inside?*"

"Well, I was upstairs just now, in bed, and—"

"Look, David, you know how excitable I am. Please. Don't go into details. Just leave out the heavy breathing, O.K.?"

"I was lying there looking up at the ceiling, and I saw these images on it, on the ceiling, like a groundhog walking across it, and—"

"I keep telling you to lay off those psychedelics. Next thing you know people will come down here and find you standing around in your birthday suit making love to your Moog."

"I wish it *were* my Moog."

"Well, monogamy isn't everything. So you found the kid's poems. I knew he wrote poetry, but I never saw any of it. You know, I barely knew these people. Charlie Mortimer—that's the first time I've ever called him Charlie—was very, ah, remote. Very straight. Right wing. N.R.A., anti-abortion, the whole shmeer. The kid was quite bright, but like something out of Charles Addams. A one-boy Graveyard School. The old man, too. Cadaverous, with a big beaked hook nose on him like that, like Dick Tracy."

Pages rustling.

"But you were going to save the piece. You were watching the groundhog walk across your ceiling. That's not bad, you know. You could get that on the *Gong*—"

"I finally figured out what it was. The shutters were closed but one of them or some of them were open just a crack. The crack was acting like the lens of a pinhole camera, projecting the image of the front lawn up onto the ceiling."

"Hmm."

"We were lying there in bed inside a *camera obscura*."

"That's great. You realize you were only born about five hundred years too late."

"That's the story of my life."

"If that groundhog of yours had walked across that ceiling just five hundred years sooner, you could have beaten Leonardo to the punch. But how is that going to bring the audience into the piece?"

"Don't you see? You can't see the image of a *camera obscura* unless you're inside it. You can't know what a magnetic field is like unless you're inside it. If you stand outside it, all you can see is the *effects* of its being there, the designs of the iron filings—"

" 'The rose in the steel dust.' "

"Right. What we've been doing is showing the audience the effects of the field, and what we want to do is bring the audience into the field itself."

"Well. I don't know. This is all pretty metaphorical. And besides, I've always liked the idea of the audience sitting in the stands and leaving the field to the players. I've never been particularly impressed with the level of play of the few *espontáneos* I've seen. By the way, the Yankee Stadium footage came back from the lab today and it's sensational. You'll love it. I brought it with me— I mean, practically, in terms of the actual mechanics of production, what are you getting at?"

"I just think we could rethink the whole concept. I don't

think it *has* to be a concert with projections, or a multi-media 'show.' That's just a *son et lumière.* The thing could happen all around the audience or in between the audience. One possibility would be an installation, with your various monitors and projections spread out all around the area, so that the audience would have to walk around to see the different parts of it. They would be inside it walking around in it."

"Hmm. Like an environment."

"Right. And one of the things I thought of when I saw those images on the ceiling is the whole idea of the focal plane. The shutters will act as a pinhole lens *all the time*, any time the rays of the sun hit them at the right angle. But unless you've got a projection screen right at the focal plane—or pretty near the focal plane—you won't see the images. You'll just see an uneven light. It was an incredible coincidence that the ceiling happened to be at just the right focal length.

"So. One idea was to have some of your projectors aimed out into the space of the room, but not in focus on any screen. I mean, they *would* be focused, but there wouldn't be any screen right there at that point, just empty space. Then, if somebody carrying a white card were to walk to that precise spot—we could mark it with tapes on the floor or something—and if they held those cards at the right angle—"

"Oh, I like it. I like it. Or they wouldn't even have to carry the cards. If they were *wearing* white, like white leotards, the image would come to a focus on them. I like this. Or they could wear nothing at all. We could project directly on the nude."

"Well, I don't know how many people in the audience are going to want to take their clothes off just to—"

"No, I mean *our* people. Then if we get some *espontáneos* to join in from the audience—with people like yourself around that shouldn't be too hard—so much the better. This is starting to sound very good."

"Another possibility would be to use video cameras *in* the installation. Then, as the audience walks around, some of what he sees on your monitors is the prepared stuff, and some is himself."

"Yeah! Only we have to hook the cameras up to a delay loop. That way, someone is checking out a monitor over here, and even though he doesn't realize it, he's being taped. But the tape doesn't play back right away. It goes into a delay loop. Then, fifteen minutes later, the same guy is over there across the room, and there—on another monitor—is him, only fifteen minutes younger."

"Perfect! See, this keeps the audience working, like you said. This audience has to be a live wire."

"Well, if he's any kind of wire at all, as long as he keeps moving he *will* be a live wire."

"I thought it was the field that had to move around the coil—"

"It amounts to the same thing, doesn't it? Oh, David, this is terrific! We're going to be immortal! I do have one small request, though."

"What's that?"

"Would you please put some clothes on?"

Danny said, "I've seen thunderstorms." Bishbash Falls had been something of a disappointment, although it had given him a chance to outswim François. "He is pretty good at tennis, but he can't hang with me in the water. It was a total face."

David said, "I don't want to make any predictions. I don't want to build it up to a big letdown. But I don't think you ought to make up your mind before you've seen it."

"Well, what's to see? A thunderstorm is a thunderstorm."

David went to refill his drink. Jane was sitting at the table in the dining room. "Did you hear that?" he asked her.

"Oh yeah. He's twelve. He isn't going to be impressed by anything—"

He clinked the ice in his glass at her and asked, "Want one?"

"Sure. I'll be out there in a minute. I just want to finish this letter to Annie."

"Tell them if they don't write us a letter or give us a call we'll disown them."

The first rumble of thunder came rolling in, tumbling into their hearing from across the river, from some other world on the far side of the horizon. Then the rains came in, with almost no wind at all. Just weight. When the full Wagnerian force of the storm took over, they turned off all the lights and sat on the porch, hearing and seeing and feeling the way it simply moved in and took command of the entire perceivable world.

"You were right," Danny said. "It *is* impressive. I've never seen *anything* like it."

David smiled, but realized also an aftertaste of sadness, a feeling of loss. He had shown the boy something he had never seen before, but that also meant he had crossed another threshold. For Danny now there would never again be another first time he experienced an electrical storm. The ones to come would all look back to this one.

As he came through the door, he recognized the music as "something familiar." After a moment or two he realized it was his own string quartet. "Why did you put this old thing on?"

"I don't know," Jane said. "I was just going through our records and there it was, and I realized I hadn't heard it in a long time. Must be years. I've always liked it.

"What do you feel like cooking? I got some lamb chops from that butcher in Rhinebeck—want to try those?"

"Sure," he said.

Listening to this quartet for him was being in a room. The room was large, a little longer than it was wide, and bare. As the thread of the quartet unwound itself, it placed things in that barrenness so that as he turned his head in one direction or another he saw the things that now gave every indication of "having been there all along"—a piece of furniture, a toy heart made out of red plastic that "beat" when you wound it up, the window that he now saw was an old-fashioned window made up of many small square panes, and on its sill was a tin box that said "Huntley & Palmers Superior Reading Biscuits," and below that, on the floor, a basket full of skeins of woolen yarn, mostly reds, browns and yellows. The room was large enough to contain the fifteen or more years since it had been composed.

He had finished it in the morning of a day when they were supposed to go to a party. They were supposed to go to a party tonight, at Anselm's. They had been living in Cambridge then, "living in sin" in a small flat on Flagg Street, around the corner from the Orson Welles Theater, and they had been invited to a party by these architect types who lived in a famous apartment in Somerville that was just two rooms wide and three stories high. The sound of thunder mixed with Wagner came down to them from on high, and he said as he looked up to the ceiling, "*Der Kinder.*"

"Yeah," Jane said. "At least it keeps them out of trouble."

She had woken up that morning in Cambridge—they were still sleeping on a mattress on the floor—feeling awful, and by noon she had thrown up a couple of times, her broad Swedish face looking pale, framed by her blonde hair, which she wore short then. He had felt an excitement since finishing the quartet around ten o'clock that morning, an indescribable anticipation that had diffused itself throughout his whole body. He had needed that party at the same time that he had known it could never possibly live up to what he wanted from it, which he could not even have put into words. She had told him to go on without her. Now he heard the main theme restated by the viola. "You're still thinking about the Riordans, huh?" he asked her.

"Yeah. I keep thinking about Annie. I know I'm not supposed to get pissed at Daniel, but I just feel like getting plastered."

"Why don't you?" he said.

Now the room in the quartet was showing him a set of orange crates, painted bright blue and filled with records, stacked against the wall under the print by Jim Dine. He had gone to the party, and found himself "attached to" by a blonde, also from California, who played second violin in an amateur chamber orchestra in Boston and still managed, as she put it, to find time to be married to a graduate student in classics. He was also at home, sick, and could David give her a ride home to Watertown? He had been seeing her at parties and concerts and poetry readings, but he could not recall exactly when they had actually met. She simply insisted that the California contingent had to stick together. Maureen Quincy. On top of the blue orange crates was a small plastic gadget that punched out labels. It was sitting on a large, flat cardboard box, which Jane had made and covered with buckram, like the binding of a fine book, and on the spine of the box she had just punched out with the little gadget: "David Lyman: Collected Works."

He hugged her now, feeling the tall, brick space of the Mortimers' kitchen around them that could not hold the four voices of the quartet which were now following each other at a walk around that other room, just walking now, but that movement charged with an anticipation that made it feel as if they were always on the verge of running. This is what being really married meant, he thought. It was a music. It created a room or a house that went with you wherever you went, penetrating the walls and roofs of any room or house you found yourself in. He had needed the party so much it could not fail him. The party was a playing field three stories high where he could run up and down all night long, while the others went on with the business of their games. He knew just enough people just well enough to keep circulating in that way, and even the dancing with Maureen was O.K., until he realized she was making her move. He was "available," she must have thought, and an edge of hard brightness came into her manner, as well as an almost proprietary readiness to touch him—on the hand, on the arm, on the elbow. At one point, talking to a law student with a glass of sherry in his hand, David heard himself being told, "You were right to come out East and see what civilization is really like." The Maureen woman standing next to him had actually given the man a raspberry, and David had smiled and thought, Well, she does have some socially redeeming value. But outside, in the painfully cold night air of January, she had slipped once on some ice and grabbed his arm and then "forgot" to let go. The intensity of her need embarrassed him, and at the same time made him feel almost protective. All the way out to Watertown she filled the interior of his car with her talk and her gestures, laughing and putting her hand on his shoulder. "He's a Casaubon," she said, finally, about Charlie, her husband, coming down a long slide of very strained humor to land on this note of seriousness and revelation that was able to include also her "Right here," as she pointed out her apartment building.

He said, "What's a Casaubon?" as he nosed the car into the curb, where his headlights glared back at him from a waist-high pile of frozen snow. It was dirty.

"A bookworm," she said.

"Well," he said, leaving the motor running.

She did not want to stop talking. "It's a half a block up the street. Could you walk me?" touching his hand. "I feel sort of funny about asking, but please—"

At the door of her building she turned toward him—she had not yet gotten her key out—and said brightly, "It feels like we've been on a date. 'Thank you for a very lovely evening.' " She reached her pursed mouth up to him to give him a playful imitation of a friendly good-night kiss. He thought, So this is what other people's marriages are like, as he felt that gritty, bitterly cold wind stumbling around his pants legs down below his coat. They were both so bundled up in woolens and gloves and mufflers that he went ahead and went along with it. It would hurt her feelings not to. But suddenly his mouth was full of her tongue and she was grabbing him through all those layers of wool and saying, "Oh, David," and telling him to come on upstairs.

"But your husband—"

"We can go in Wade and Christie's apartment. They're in California. I've got the key."

"Look," he started to say. And the next thing he knew he was walking up a short, dimly lit flight of hallway stairs, and then he was pinned against the inside of the front door of "Wade and Christie's" apartment, looking down the long straight hall toward the kitchen. He could feel her mittens at the back of his neck and the thick, woolen weight of her against his chest. Looking over her shoulder down that hallway was like standing in a square tube or conduit that was ready, waiting to be filled, to convey some fluid. It was as cold in this apartment as it was outside. "Look," he started to say, his breath steaming. On his right a door opened onto a living room where a fireplace had been converted to a gas heater. Above the mantel was a mirror that showed him the blank yellow of a wall in the middle of which floated two pictures that might have been Utrillos.

"Why can't he be like you," she was saying, "alive."

"Look," he kept starting to say. The sliding doors between the living room and the bedroom were open.

"This is crazy," he said. He was standing inside somebody else's apartment, somebody named "Wade and Christie," whom he had never met, and in this place where these total strangers had their lives he was being forcibly seduced by the wife of somebody else whom he had never met, someone named Charlie, a scholar of classics, a bookworm who might this very minute be upstairs, just a few feet above his head, trying to translate Cicero but being constantly distracted from it by paranoid fantasies of this sexy blonde from California he had married two years ago because she seemed so strong and who was now lying face down with her rear end draped over the armrest of the sofa in the living room of the Chelsea Street apartment of an Arab architecture student whom she had picked up at this party to which Charlie had insisted she go without him, and who was now standing between her spread haunches, his pants spilled around his shoes on the red kilim, fucking her in the ass while she hugged the cushion of the couch to her bosom and whimpered like a dog with pleasure. How many lives could he stand in the middle of at once?

"I know it's crazy," she said. "It's mad and it's bad and it's sad, but oh it's sweet! When I saw you conducting the Mahler I said to myself, 'God, what an animal!' "

Through the sliding doors, two rooms away, he could see a lamp on the nightstand next to the bed, a plastic lamp in the shape of a goose. It didn't look like a child's bedroom.

"This is awful," he said now to Jane. "Do you mind if I turn this off? I'd like to hear some Cecil Taylor."

"Sure," Jane said.

The party at Anselm's was a little dull. In the car on the way back, Jane had said, "The men are all being *very subtle* about going into the other room to snort cocaine, and the wives are still telling jokes about their pet rocks. I love Anselm as much as you do, but those video people from the city are the *pits.* Competitive? Dog-eat-dog? And this is supposed to be the capital of civilization."

When they got back to the house, Danny and François met them at the front door to tell them about their adventure. They had gone outside to shoot off the last of the firecrackers, and then realized they had locked themselves out. François gave him a boost up onto the roof of the back porch, and Danny had gotten in through the window. The slanted roof was really slippery, but getting in the window was easy. Anyway, while he was up there he found the beehive, and in snooping around it had actually riled up the bees and he got stung once, on the back of his head. "But now we know the source of the bees," Danny said.

"Wonderful," Jane said. "We solve all the mysteries just before we have to leave the house."

The next morning, packing the car, he looked down the driveway at the back porch and the metal roof that sloped up from it to the window Danny had climbed in. It was a foolish thing to do. They could have hung around outside for a couple of hours until the party was over, or they could have walked up to the college to use a phone to call Anselm's. The sun that was shining now on the window and the roof and the driveway would be shining like this too, even if the night before had brought them some bad luck.

They were an endless source of worry. If it was not falling out a window or off a roof, it was getting run over. By a beer truck. In Harrisburg, Pennsylvania. You worried about them falling off the roof trying to get in the second-story window, and then you worried about them getting too good at getting into second-story windows and ending up like that kid who had ripped them off. Albert Boone. The police had spotted the license number and David had picked him out of a lineup the day before they left California. Twenty-five years old. A six-time loser. He had never held a job and didn't even have a social security number. His arrests were handled routinely by the Career Criminal Division of the D.A.'s office. He was a tall, good-looking kid who was built like a basketball player and had a light of something in his eye that David admired, the way he'd admired the washcloth over the license plate. At the lineup he had *strolled* into the booth wearing his white prison jump suit with an arrogant, animal grace, and had blown a kiss at the unseen audience on the other side of the glass. David had wanted to applaud him then. But the romance of the gesture had to be followed by the reality of lived time. Another waste.

And the Mortimer boy, that Mozart-Rimbaud-Norbert-Wiener. At age twelve, crushed under the wheels of a beer truck. And the father, "a brilliant and tireless researcher," the dust jacket of his book had said, indicted for "murder one," as the TV cops would put it, before his son had even discovered girls. Danny, too, for all his swagger, only *looked* at the music-camp girls. He was still at that age, in spite of his size and his apparent maturity, when he did his wrestling and grab-assing with François and other boys. Jane had wanted to say something to him about the girls from the music camp, but what the hell. Soon enough. Soon enough David would find copies of *Playboy* in the back of his drawer, or a handful of those little foil-wrapped rubbers. Plenty soon enough the child in him would die. And even though it had to happen and everyone wanted it to, everyone but the boy himself felt the loss. Maybe the boy felt it too. Maybe later.

The path entered the woods by a "door" in this wall of trees, and even for twenty-five feet or so inside these woods you could still look back and see the house through a couple of "windows" in the branches of the outermost trees. He hefted the shoulder strap of the Uher and began to adjust the headphones over his ears. As he walked, he held the long barrel of Anselm's mike out in front of him. He felt like a fisherman or a hunter. And so I am, he thought. He aimed the mike at a bird on a branch some fifty feet away, and heard its chirps come through the headphones with piercing clarity. But two feet in either direction and the bird faded out.

It felt strange to be carrying this extremely sophisticated equipment in here, in the ferny green shade of these woods. But that's what it took, he thought, to bring it back alive. He saw a small clump of Indian pipe, a tall, slender fungus that looked like a translucent waxy-white flower in the shape of a pipe. They were growing out of an old dead log. And here were some tiger lilies in a row along the path, just like the ones along the driveway.

As he walked along, he swept the barrel of the microphone slowly this way and that in front of him, covering the woods on either side of the path. He had the machine set on pause, and was ready to take whatever the woods might give him—if he liked it. So far there had been nothing but the generalized background noise of leaves and branches moving in a small forest. He wasn't looking (or listening) for anything in particular; he only knew that if something were to come in now over the headphones, something O.K., he would recognize it as such. Like the voice of the house, that imperfect fifth.

On one sweep he heard something crunching, repeatedly, and aimed the mike. It was something chewing. Some small animal, a squirrel or a groundhog. He could hear when the animal was biting a piece off or when he was chewing a mouthful. It was chewing on something chunky and crisp, like a root or a nut. This was charming. This was an aspect of the woods he had never really thought of. Of course he knew about it all along, but somehow he never consciously thought of the woods as a place where animals *ate*. Sometimes each other.

He passed the clearing where somebody had set up a small "living room"—couch, easy chair, coffee table, a floor lamp. Even a picture hanging from the ivy-covered trunk of a beech tree. It had been there a long time and had taken a beating. When they had first walked past it, David had thought the Mortimers had put it there, but François told him, "No. Acidheads from the college. They would come here to trip out."

François said, "There. See him?" very softly.

He had heard François's voice over the headphones. He looked in the direction the microphone was aiming. The bushes were over his head. He could not see anything but the mass of twiggy, leafy branches and trunks of trees, bushes with long, pointed leaves and trees with broad, flat ones, heavy limbs that moved horizontally and masses of thinner branches ramifying in all directions, most of them masked by the green of the leaves. They formed a "wall" he knew would be many feet thick. Somewhere in the middle of that world, unseen, François said again, "There," and David heard the electronic reconstruction of his voice come through the headphones over his ears. Then he heard Danny's voice saying, very coldly, as if preoccupied with something else, "Yeah, I see him. I'll waste the fucker's ass." And then a *twang!* so loud it startled him. He flinched and found himself paralyzed with horror at what he was hearing. When he flinched, the barrel of the mike jerked to the right, and through the headphones he heard a *thwup*, like a thick body being struck. And then the screams. Someone was dying. Something had struck him—no, her: it sounded like a woman—and she was bleeding and dying, and the screams were enraged and helpless. They were the whole life of that woman giving itself to the speaking of this one panicky bellowing "No!" to death. Trying to say "No" to death. The sound went straight through his breast. The screams were now even gargling, as if the blood had entered the throat, entered the voice itself.

He dropped the mike and tore off the headphones, bellowing himself, "No!" He leaped straight ahead into the wall of bushes and trees, clawing aside the branches and screaming, "*No*," ripping the tape recorder off himself as he went, the branches and twigs scratching at his face and his arms, poking at his eyes and gouging into the skin of his arms and hands, seeing for a long moment only the greens and browns of the bushes he was tearing through.

They had been too scared of his own cry to run. When he broke through the brush, he saw them huddled against a tree and when he turned toward them Danny actually cowered, raising his arm to ward off a blow. He could still hear the screams, off to his right, where he could see something small and gray-brown thrashing convulsively in a clump of ferns. It was a groundhog.

They had shot a groundhog, which was now lying on its back, its paws fiercely punching the air above it and screaming. The arrow had gone in at one shoulder and about one inch of the steel tip had pushed its way through the thick part of the animal's body to emerge just below the other leg. Blood was coming out of its mouth and splattering the ferns, which joined in the thrashing.

He picked up a branch about the size of a baseball bat and took careful aim. He brought the branch out and down in one sweeping arc into the animal's head. It actually made a dent in the head and felt like hitting a rolled-up rug except that a rolled-up rug is the same all the way through, and this animal was made up of a thin layer of fur and skin laid over the bare skull whose brittleness had tried to protect the soft, vulnerable organ of the brain but had failed under the impact of the wood David had slammed into it with all his weight.

The animal lay still. The ferns, in another moment, were also still, and the screaming had stopped. The groundhog was quiet. The clearing was silent. The forest continued to produce the general background noise of leaves and branches moving against one another in the breezes. He turned to face the children, but they were not there anymore.

As he came in from the bright sunlight, he heard the TV going. Monster-movie music, complete with crashing thunder and driving rain. François had come back from Nova Scotia two days before, and David was starting to feel a little pissed to see a pattern developing—sitting around the house all day watching the boob tube instead of being outdoors doing something wholesome—swimming or tennis or whatever. All this incredible weather going to waste.

There was something strange about that music, but before he could figure out what it was he was inside the TV room, which was empty. The set was on, but it was a quiz show, whose host smiled maniacally as quiz-show music enveloped him in its hectic brasses. The blaring stopped abruptly as the screen cut to a young woman in a tennis dress sitting in front of the mirror of a dressing table. She was talking about Tampax and David could still hear the monster-movie music and the thunder. It was coming from over his head, upstairs. Snippets of Wagner and Richard Strauss.

He switched off the set and walked upstairs, thinking, Why can't they just listen to Billy Joel or Blue Oyster Cult, like normal kids? When he looked into Danny's room it was empty too, and he remembered for one panicky moment that the front door had been standing wide open when he came in. What if someone had come in while the kids were in here by themselves?

The door to Danny's closet stood a little ajar, and the music was coming from there.

"Danny?" he called out. "François?"

A large pile of clothes was lying rumpled on the guest bed. He moved into the room and went to the closet. He pulled open the door. In the back wall of the closet, three feet in front of him, was another door, also ajar, that he had never known was there. The music floated out through the opening on a wave of thunder that actually drowned out his call: "Danny? François?"

When he opened the door, he saw a steep flight of stairs going up into a dimness that was now filled with the pulsations and electronic screechings of the music. A flash like lightning suddenly etched the whole ceiling above the stairwell into blinding, momentary being. Underneath the music was a steady hum, and with it some other noises, like some sort of machine, repeating themselves at regular intervals. The noises were something small, like a sewing machine, and then over that steady humming there were these other regular clunks and bumps. Now the thunder rolled down the stairs into him, actually jolting him.

As he climbed the stairs, more and more of the room came into view above the stairwell, the joists of the ceiling and the studs of the walls lit by flashes of lightning. Whatever was making that humming mechanical noise was something that was spread out, that took up the whole area, something that must have been on top of the tables he now saw set up on sawhorses, the tables set together with their edges flush so that they made a sort of second floor up there, about waist high. A flash of lightning lit up with painful clarity the model trains and tracks and towns and mountains that covered these tables.

From behind him Danny's voice cried, "Hi, Dad!"

He turned, and there, in the dimness of the room, behind some sort of panel, were Danny and François.

"What the hell?" David said.

Danny said something he could not hear. The boy was smiling. Now he turned toward François. The music softened for a moment and David heard him say, "Turn it down for a minute," and abruptly the level of the music dropped to almost nothing.

David could feel his own face go wide in a smile as he asked, gesturing to take in the whole room, "What *is* this?"

"Isn't it neat!" Danny said. He said something else, but at that moment an enormous clap of thunder stunned David from behind. Danny turned to François and made a scissors motion with his fingers. François reached over to a switch on the panel. Silence, except for the hums and clicks of the trains as they toiled along their tracks, even their speed scaled down to size.

"What is this?" David was saying. "Whose is this? Has this been here all along?"

"Isn't it incredible! François showed it to me. We're not breaking anything. François knows how to work it."

Just then François turned up a rheostat and the lights came up and the room grew into visibility all around him.

David was surprised to see how small it really was. Every available inch of space was taken up by the model railroad, whose tracks spread out in all directions, going even beyond the room, into the sort of crawl space between the roof of the house and the ceilings of the various rooms. Five or six different trains must have been going at once, some of them puffing smoke, others sleek modern diesels with immense long lines of cars stretched out behind them around curves and into and out of tunnels, into the mountains and around the beams that supported the roof. Along the tracks billboards advertised whiskey and cigarettes, as well as a movie version ("All of Its Thundering Power Comes to the Screen!") of R. Charles Mortimer, Jr.'s "Beloved Best-Seller, *Gens and Justice: The Law in Early Roman Civilization.*" Forests of individual trees covered vast stretches of the land. In the mountains a logging camp was in full operation; a log sluice filled with running water was carrying scale-model logs to a pond where they were loaded onto flatcars. Behind a beam David saw a quarry, and off in a far corner of the crawl space above the ceiling of his own bedroom a factory was working, some of its windows lit, some of its chimneys emitting wispy puffs of smoke. Against the far wall the wide gray Hudson flowed. The sharp smell he had noticed was creosote. Even the ties of the railroad tracks had been brushed with creosote. Near the control panel was a town with a depot and a post office, houses with backyards with real laundry hanging from the clotheslines. *Coming Home* was playing at the local movie theater. The hoops on the backboards in the playground of the school had real nets on them. And in this entire world they were the only three people.

François had known about it all along, but it had been more than a year since he had been up here. He used to come up here with the Mortimer boy "all the time" before, before that happened. He had learned how to work the panel then, although he still forgot a switch open someplace and derailed a train from time to time.

Mr. Mortimer and his son had designed and built it. It had thirty-six different engines, and a railroad yard ("over there") where the engines could be made up into trains out of the hundreds and hundreds of cars. They had also installed six tape decks in the little control booth and about twenty-four speakers either against the roof or under the table the trains ran on. Four different projectors and several batteries of strobe lights took care of the lighting effects. It could do the electrical storm, and it could do a sunset and a dawn. It could also do a sort of psychedelic light show or a 1920s flickering train movie, or it could just do train music.

"Did the boy write the music?"

"Yeah. He didn't have to go to school or anything. He was smarter than all the teachers. He made up the music on his synthesizer downstairs. Some days he wouldn't do anything all day long but make up music on that thing. He would set it up so that you couldn't even hear it; the music would go straight from the machine, through the wires and into the tape."

It was pretty schlocky music, David thought, sort of a cross between Ferde Grofé and Miklos Rózsa, but for twelve years old! Of course it could also have been tongue-in-cheek. He looked at Danny, who was sitting there just shaking his head. He saw David looking at him, and said, "Hey, Dad, did you see this?"

It was a house at the edge of the woods. It had what appeared to be a whole collection of roofs, all covered with something like tinfoil, all sloping at different angles. Looking through the window of the TV room, David saw a scale-model TV set, which must have had a tiny blue bulb in it: its eerie light filled the little room.

"This is incredible," David said.

"Look at this," Danny said, and pushed a button on the control panel. The entire top story of the house lifted up on a shaft that came up through the center of the model, revealing the rooms of the first floor—the dining room, set for a formal dinner with several goblets at each place setting, the kitchen with the badly installed dishwasher. On the windowsill and on the floor below it were tiny dots representing the bees.

He looked at Danny, who was saying, "That isn't even what I wanted to show you. I pushed the wrong button."

Now the first floor of the house rose up on the shaft and closed against the bottom of the raised second story. In the cellar of the house was an exact replica of the studio. The tape reels could be moved with your fingers. The synthesizer stood in its appropriate corner. On the top shelf of the storage unit was a tiny black binder.

David looked up at Danny, shaking his head. Danny had an enormous grin on his face. He pushed another button and David saw the two stories of the house lower themselves with a click over the cellar. But the roof stayed where it was, allowing him to see into the bedrooms, the upstairs bathroom, even the closets. There was the secret door in the back of the closet in Danny's room, there were the steps that led up to the secret room where the trains were.

He looked at Danny, realizing that his own mouth was open. His son was loving this, sitting in the little booth with his hands on the controls. The boy pointed at the house, and David looked again. Now the roofs all rose a little higher while the walls that had connected them with the second floor lowered themselves. He was looking into the room he was standing in, the train set spread out on all the tables, the tracks running far beyond the room itself into the spaces above the ceilings of the other rooms, the tracks sometimes running past miles of forest. On the far side of the room was the wide gray Hudson. Near the control panel was a white house at the edge of the woods.

On the floor inside the little control booth, under the built-in bench the boys were sitting on, was a cardboard box. Danny was much too excited, but François remembered to ask David if he would like to "work the panel." The only way to get into the control booth was to crawl under the table, and that was when David saw the box.

"What's this?" he asked. But neither of the boys had noticed it. Inside the cardboard carton marked "Bose" were several dozen boxed reels of tape, a tall, stiff-bound notebook marked "Ledger" on the spine, half filled with irregularly dated journal entries, some in French and German and some in what looked like a homemade shorthand, and various rubber-banded bundles of papers and letters, including one batch of letters from an attorney in Harrisburg. From the papers in the box, the poems he had found in the studio, the house itself and François and his dad, David was able to reconstitute the whole story.

R. Charles Mortimer, Jr., inherited his father's prudence and single-mindedness. R. Charles Mortimer had been born in Orono, Maine, and studied physics at Bowdoin College, and then at M.I.T. He did not like Cambridge, and missed his family painfully while he was there, but he had been told that M.I.T. was the place to go if you wanted to insure yourself a good job, and so he stayed and persevered. Physics to him, as he plied his slide rule in his dormitory room or in the library, was a soaring music that bathed the entire universe and every piece thereof in its splendor. It explained, it predicted, it controlled. It was all vast principles and immutable laws, which even as they dispelled mystery, grew in elegance and in scope until they had become mysterious themselves. Like a music that flowed, enfolding the room in a glow and creating a room, an architecture of its own, physics erected a structure as massive and delicate as a cathedral.

When he completed his degree, he took a job with an engineering firm in New York City. He had never been to New York before and did not like it when he got there—it was so big and dirty, and the people were even more brusque than in Cambridge—but it was a bird in the hand. The salary was a handsome one and he could begin immediately to repay his father the cost of his education, as he had agreed. It was the first job he was offered, and it was the only company for which he ever worked. After he had settled his obligation to his father, he began to allow himself small pleasures, like accepting the invitations to go to Staten Island on picnics with the family who lived across the hall, whose eldest daughter he eventually married.

In his work he quickly calculated stresses and tolerances to seven or eight decimal places, and this efficient, meticulous thoroughness was not lost on his employers, who gladly advanced him in salary and rank over the years; the advances brought with them, as his wife suggested, moves to larger apartments, and eventually to a house near Stony Brook, in whose study he sat most evenings after dinner, painting miniatures with the aid of a magnifying glass set in a metal stand that he had designed himself and then had fabricated at a machine shop which a co-worker had told him about. He brought the same attention to detail, the same quiet assumption—that error-free work was within the unhurried grasp of everyone—to his dealings with people. Errors, either of commission or omission, were not heinous sins, but simply part of the process one got through, patiently. You corrected them and went ahead. If someone else made them you pointed them out, and if he made them again you replaced him, though that equation too sometimes worked out differently, when, for the sake of economy, it was necessary to tolerate a measure of inefficiency.

It was not long before he found himself doing less and less engineering, which he had never liked—it was all plodding routine, worked out according to recipes, with the air conditioner for accompaniment—and more overseeing of the work of the others—draftsmen, and eventually even senior engineers—which he also did not enjoy. He learned their names because it was necessary.

The one time his son saw his office was from a suite in a hotel on the downtown side of Fifty-third Street. He was sitting in one room of this suite, waiting to be interviewed in the adjoining room for his first (and only) appointment as Assistant Professor of History. The other candidate for the position lacked poise. Before the interview, R. Charles Mortimer, Jr., had walked up Seventh Avenue knowing that this was where his father worked, and feeling a thrill of secret knowledge as he passed the doorway that bore the address of his father's building. It wasn't that he had disobeyed his father; rather, he had done and was continuing to do something that did not bear out his father's expectations or assumptions, and so there was a thrill of the vaguely forbidden, or at least disapproved-of, to the simple fact of his walking up this sidewalk, *like* these other people and yet *un*like them. He had in his mind an image of his father some forty-three stories above him, working and perhaps thinking of the street outside and below the windows of his office. But his father's conception of that street did not include himself walking there, walking there "looking for a job."

His father had assumed that he would enter a profession, and that it would be a profession that rewarded endeavor and competence adequately. He had listened to his son with a vague distaste when the lad had come into his study to tell him that he did not want to practice law but to study it, especially its origins and its development in history. Mr. Mortimer was still holding his fine-pointed brush. He looked from his son's face to the picture he was rendering in miniature, a photomicrograph of zinc, whose magnified image swam in the lens in front of him as he moved his head. I made a mistake, he thought.

"This is a mistake," he said.

"It's what I want to do," his son said.

"Nothing can be done."

The decision had changed forever the quality of the distance between them. R. Charles Mortimer, Jr., now looked out the window into the office in the building across the street and two stories down. His father was standing in front of a drafting table. A man in shirt sleeves was standing next to him looking down at the same drawing. The man was talking, and from time to time he pointed to something in the drawing. His father moved only once, pointing with his pencil at a corner of the drawing, which the man rolled up and took away. That was what his father "did." When people asked him what his father did—kids at school or at college, or the woman whom he had already asked to marry him (after he had gotten established)—he told them he was an engineer or that he worked for an engineering firm. This is what it meant. Just then Professor Hartpence opened the door and asked him to come in.

The one other imprudent thing he did was to marry Allison before his position at the college was secure, before he had been given tenure, but in the end no harm had come from that. She had introduced herself to him at the tennis club that his mother had insisted he join because it would give him poise and confidence, and so he played tennis. On some of their dates he discovered that he was very passionate in his desires. He thought about these passions guiltily after he got home from these encounters, wondering where such intensities of feeling might lead him.

He asked her to marry him in the summer before his junior year at college, and they both settled down into something that felt like a zone of being together that lasted more than seven years. On the first anniversary of their engagement he "almost lost his head" again as he was kissing her good night, and from then on he was more careful on their regular dates and especially watchful when they went out to celebrate their anniversary.

When he told Allison about the letter informing him of his appointment, she said, "I think we ought to get married this summer."

"But I don't even have a house."

"We'll get an apartment."

"But I don't even know if there *are* any apartments up there." He had imagined himself sharing a dorm room with some other new assistant professor, as he had always shared dorm rooms with a different student every year.

"We'll go look." And the next weekend they took the train from Grand Central. Two hours later they were walking along the tree-lined side streets of Rhinebeck looking for FOR RENT signs, and eventually they made arrangements with a German woman who ran a boardinghouse. His father said nothing, but kissed the bride and sent them on a honeymoon to Europe. After that they settled down to being a professor and a professor's wife, living in two adjoining rooms of Mrs. Rückert's boardinghouse. It was in those days that he formed the habit of doing his research and writing at the college.

He learned the students' names as well as he could, because it was necessary. But as soon as his classroom or class-connected duties were done he would close his office door behind the last student and either open the book at the place marked as it sat on its little stand beside his desk, or put his coat on over his cardigan sweater and walk to his carrel in the library. There, in the cool underground catacomb-like stacks, he would pore over photocopies of papyrus documents and slowly, painstakingly, come to his conclusions.

Sometimes, from the desk of his carrel two stories underground, he would look down the long, narrow hallway between the stacks, lit only every twenty feet or so by a bulb in a cage that jutted down from the ceiling like the root of a plant. The books themselves were so slender, about a hundred of them would fit on a shelf. It gave him an odd excitement to be here, under the ground, in touch in this way with the lives of other people. Sometimes (but only after he had completed the work he had allotted himself for that day) he would stroll down the aisles stacked from floor to ceiling with shelves of books, touching and randomly taking down books and reading their dedications and acknowledgments: "To my mentor and guide," "To my students in the seminar on," "To my colleagues ——and ——, without whom," "To my friend, Professor ——," "To my brother," "But most of all to my wife, who typed these pages with such painstaking care"; "who compiled the index"; "who brought me cups of coffee all that dark autumn in Reykjavik"; "who brought me lemonade on the terrace of our little pension in Firenze"; "who stroked the cat and stoked the fire."

He would get something of this feeling of excitement when he walked down the hall at the college and saw, through the open crack of an office door, the back of one of his colleagues bent over a typewriter. But there the totality of the effect was interfered with by the actual presence of the man, grossly affirming its rights there. Once he walked into Hartpence's office to ask him something, and to his surprise, though the door was ajar, Hartpence was not there. Then he had something of the feeling he'd had looking out the window across Fifty-third Street and down two stories into his father's office.

The dedications and acknowledgments in the library were not the same excitement, but they pointed to the same feeling. They were the effect of something he could know only in this way. When these authors had written them, who had they thought would read them? And where? How could they have imagined *him* standing here in this underground dim silence, holding the book open to this early page, so near the outside cover of the book and yet so much closer to its living center than the text? The feeling he got then was satisfying, even though he knew that he was at those moments being in a place that stubbornly continued to be empty in spite of his being in it. He had dedicated his own book to his father, and in the acknowledgments had closed by saying, "and especially to my wife, whose loving patience sustained me as I worked." He had hoped she would not ask him about it, and she did not.

She played tennis when she could, and when her father died and her mother moved back to Harrisburg with her half of the inheritance, she bought a membership in the local tennis club for the two of them. Then she bought the white house at the edge of the woods. The sets of books were there when they moved in. Allison put a baby grand in the living room and had the kitchen redone. She resumed her piano lessons and started cooking lessons. She also started the women's chorus, and brought a kind of ferocity into their lovemaking, which was only the latest of the things about his wife that he neither expected nor understood.

One morning at breakfast she looked at him across the table and said, "I'm pregnant." He thought her tone was funny, but it was just another of the things about her he did not understand, in a way similar to this thing now going on inside her that was called "being pregnant." She was now like one of those books or one of those offices when the professor had gone off and left the door open.

She was the place of a life that excluded him. At the window a couple of bees were buzzing to get out.

His son was born and his father died in the same week, and he found himself independently wealthy at a time when he was too preoccupied to take any advantage of that wealth. He continued to work because he always had, and it had never occurred to him to do anything else. The son was a lot to preoccupy him.

As a baby he did not smile. When he looked at people, especially children, it made them uncomfortable because he never blinked. This is not unusual in babies, but in Mortimer's son it was combined with a look of such penetration that he made people wonder if this wasn't something different from "regular alertness." He looked as if he were simply biding his time till he could learn the language these fools spoke, in order to tell them—without anger or resentment, and only a trace of impatience—what fools they were. He almost never cried. He never went through a babbling stage. He never babbled, and so far as anyone could tell, he never made any mistake. He spoke his first word at six months: "Up," meaning "Pick me up." By a year, he was having conversations, and he first read Dr. Seuss, whose books he very quickly left behind, before he was two.

Nobody was quite sure when he discovered the piano or the woods. Several times Allison heard the piano, but she thought it was Valerie, the student who was sitting him in the afternoons, and Valerie of course thought it was Allison. Then one day Valerie walked into the kitchen just as Allison was coming in the back door wearing her tennis outfit and looking a little sweaty, and the two women chatted for a moment before Allison asked where the boy was. He was in the living room playing the notes of the Liszt he had heard his mother play. He was five then, and within the year he was playing not only better than she but better than Myra Brubaker, who had studied at Juilliard.

There was never any question of his going to school. By the time he was old enough for the first grade he knew more than the teachers, and by the time he was seven he was composing music, most of it lugubrious, which was not surprising, given the boy's characteristic manner. He had not smiled as a baby, and did not often smile as a boy. He was not pouty or bad-tempered, just somewhere else. As often as not it was the woods.

It began when he was around five that his mother or the sitter would take her eyes off him for a second and he would simply not be there the next time she looked. Panic and lots of calling for him throughout the house and into the green mass of trees and leaves and branches that was the woods, and after half an hour or so they would hear him answering, "Yes?"—calmly, as if he could not fathom or even acknowledge their upset.

"Where have you *been?*"

"Next door."

One day Valerie followed him, keeping just out of sight and ducking behind a tree whenever he turned around, till he looked back and called out to her: "It's O.K. You don't have to hide." But when she came up and walked beside him he had no more to say to her than he ever did.

"What are you doing?"

"Looking."

"Looking for what?"

"Oh, things."

And just then he found one, and bent down to examine a large, flat orange mushroom growing horizontally out of a fallen log.

"What's that?" she said.

He gave her its Latin name and added, "Poisonous."

"Oooh," she said. "I wouldn't eat *any* of them."

"It's beautiful," he said, "and the good ones are delicious."

At other times he would stand very still, his arms held out a little from his body, his head bent forward slightly, cocking it from side to side.

"What is it?" Valerie asked.

Without moving his head, his eyes clicked over to give her a look that said he *would* give her an answer but right now he did not want to break his concentration.

"The woods," he said.

He never stopped walking in the woods, even after he had persuaded his father to let him build the studio down in the cellar. Later his father would go with him on these walks, go everywhere with him, and they would build the railroad room together. But at first when he went "next door" alone he brought back specimens—butterflies and crickets, moths and mayflies, squirrels and chipmunks and once even a groundhog—all dead. He would explain slowly and gravely what they were, their genus and their species and so forth, and how they died and how he had found them. Then he would allow them to be taken away and disposed of.

He never formed any attachment to these things or to any favorite blankets or stuffed animals or toys or to any imaginary playmates or to any actual playmates. One day, a year or so before the end, he met François on a path in the woods and brought him home to play with the railroad. But Mr. Mortimer suspected that the boy had heard him and his wife talking the night before about their concern that he had no playmates. François was introduced, formally, as the two boys stood in the living room. "This is my friend François. May we go up and play with the train together?"

Up until then he had divided his time between the studio, the woods, the train room and his own carrel in the stacks of the library at the college—his father had made the arrangements for the carrel. At first he went to the library almost every weekday, always pausing as he went past the chapel to look down into the cellar of the Barstow vault. But later he went only every other day, and finally, toward the last, almost not at all.

One day R. Charles Mortimer, Jr., looked into his son's eyes, and the change that came over him from that day forward was remarked on by the entire community. He did not change his manner, which remained restrained and proper, but he began, for the first time in anyone's certain knowledge, to spend time with another person. Mortimer had never been a hermit, and he and his wife regularly invited his colleagues to dinner or accepted their invitations to dinner. They went to all the required social functions and, generally, blended into the background, where people assumed they enjoyed themselves. They appeared in the audience at concerts and plays, and Mr. Mortimer was always there when the women's chorus performed. He even played tennis, with a somewhat mechanical style, and was always gracious in defeat, and even, on those occasions, in victory. Mr. Mortimer played tennis because people expected him to play, and invited him to play, so he did, and when the game was over they expected him to behave in a certain way, so he did, although he could never imitate their jocose, bantering manner. But no one had ever known him to choose to spend a lot of time with any other person.

On the day when Valerie had gone with the boy into the woods where he had shown her the poisonous mushroom, the two of them had come home to find Mr. Mortimer just getting back from the college.

"We've been to the forest!" Valerie told him excitedly.

"Oh," Mr. Mortimer said. He looked a little surprised to see his son, as if he had been preoccupied with something. He looked down at the boy, who was still holding Valerie's hand and looking straight ahead. Mr. Mortimer had never been in the woods himself, but he knew that he was expected now to say something, to ask something of the boy, as if it had been he who had just spoken.

"And what did you find there?" he asked, looking down at the top of the boy's head and the bridge of his nose.

The boy turned up eyes of a blue so pale and transparent they seemed to go on forever. Mr. Mortimer knelt down to look more closely into his son's face, and the boy said something he recognized as Latin, but did not understand.

"That's the Latin name of a poisonous mushroom I found and showed to Valerie," he said. "It's a beautiful mottled orange and kills within twenty-four hours. There isn't any antidote that anyone knows of." He continued to look into his father's eyes. And as he looked into his son's eyes the father changed. There was something in that look that he recognized.

He knew that Valerie had not suggested going into the woods or she would have said, "I took him into the woods and we found—" Besides, the boy had been disappearing "next door" for several weeks now, as his wife had told him. So he would have gone on his own, and Valerie would have followed, trying to keep out of sight and not doing a very good job of it, certainly not good enough to fool the boy, who would then have invited her to come along. He had nothing to hide. He had invited her into the world of his life, and she had thought she was "watching" him, even after he had shown her what he knew that she did not know. Even now she had said, "We've been to the forest," as if it had been her idea, and the boy was perfectly willing to tolerate what she did, since ultimately it did not concern him in any way. Tolerating this sort of thing made people like Valerie happy, put them at their ease, and then they left you alone more.

The father knew this, and while he looked into his son's eyes as the two young people stood and he knelt like this on one knee there in the driveway, and the afternoon sun shone down on them and raised the smell of spilled oil to his nose, he could see the boy walking "alone" into the woods, down that long first path that proceeded through the forest like a straight, wide corridor into its depths. Within a hundred feet this corridor was intersected by another. The boy, as his father saw him from behind, was walking (he had never toddled) straight and upright, with his hands in his pockets, sweeping his head from side to side across the path and the woods on either side. Then he stopped, turned halfway around and called out to Valerie, "It's O.K. You don't have to hide."

It was a life that had been going on while he was in his carrel at the library, looking down that long silent hall lined with books and lit only intermittently by those bulbs. This was something completely different from the dedications or the acknowledgments, different even from the offices he had looked into when their doors were open a crack or when he had stood in Hartpence's office alone, unseen, feeling the presence of the woman who had knit this hideous afghan and who had no conception of his standing here in the zone that it helped to create, this room that was a room in someone's life.

Finally Mr. Mortimer said, "I would like to see that mushroom. Could you show it to me?"

"Yes," the boy said.

He stood up and then he and his son walked off up the driveway toward the woods, hearing Valerie say, "I'll just go then, O.K.?"

"Certainly," Mr. Mortimer said.

From that day on he spent every available moment—when he was not teaching or involved in some other duty connected with his position or working in his carrel in the library—with his son. He did not take the boy to baseball games or to the zoo or the aquarium or to the Empire State Building or the Statue of Liberty or to the amusement parks in the Berkshires. He only walked with the boy in the woods or to and from the college when they both went to use the library, or he would drive with him into town on one errand or another. They never spoke much, and the only time he ever suggested something—a circus in Poughkeepsie, which he'd seen advertised on a telephone pole outside the butcher shop—the boy said he did not think he would be much interested, and then added, "Thank you."

When the boy asked him if they could convert the cellar into a recording and composing studio, he had asked how much it would cost, and the boy had shown him six pages of figures itemizing the cost of every piece of equipment and the store or discount outlet where it could be gotten at that price, and the estimated cost of labor and materials for the conversion process itself. Mr. Mortimer agreed, and they had the construction work done while the family made its annual pilgrimage to Harrisburg, to visit for three weeks with his mother-in-law, who always asked what sorts of sports, *athletic* things, her grandson was participating in.

But when the studio was completed and the boy began to spend more and more time in there by himself, wearing headphones and composing, on his various machines and tape recorders and keyboards, music that, to the best of Mr. Mortimer's knowledge, nobody ever heard, he said to himself, "I made a mistake." That was when he had begun investigating the advertising brochures for various instruments that could be played "by anyone," and finally bought the chord organ, thinking that perhaps that way they could all—his wife and the boy and he—play together. Perhaps they would even be able to play some of the pieces the boy said he was writing. He asked his son then, "Can we hear some of the music you're writing?"

"We?" his son asked him.

"Your mother and I. We know you're working very hard on these pieces, and we would like to hear some of them. One at least."

The boy looked up at him with those limpid blue eyes, and then looked away, up into a corner of the studio ceiling, above the top shelf of the storage unit. He answered then without looking at his father. "Mother is not at all interested or she would have asked me by now, but if you would like to hear some I think I could start you off with the songs I was doing last year. Or would you like to hear the stuff I'm doing now?"

His father would be pleased to hear the songs from last year, and the boy walked over to the storage unit, picked out a boxed tape reel and handed it to him without a word. When he listened to them later, Mr. Mortimer was struck first of all by the fact that these songs had no words, and then by the dirge-like, lugubrious heaviness of the "music," which was made up of sounds only some of which he recognized, like the piano. He had threaded the tape and turned on the machine with some anxiety, and was relieved that the songs, though harsh and in some places discordant and overall almost unbearably sad, were still immediately recognizable as music. He had been anxious as he pressed the playback button because, when his son had handed him the tape, Mr. Mortimer had said, "Thank you, but I would also like to hear the music you are writing now, if I could," and the boy had wordlessly rewound the tape and then, with his hand on the control panel, had looked at him for the first time since being asked about his music.

"This switches the sound," he explained, turning a knob, "from the headphones to the speakers," and he pointed up to the enormous speakers up near the ceiling. Then, with his hand on the play button, he gave his father a long, steady look that was the equivalent of a shrug of the shoulders, and pressed the button. Immediately Mr. Mortimer felt his flesh creep. Something was skittering around on a series of taut strings, like mice over dead leaves. It was a pattern that made a texture that covered the walls and floor and ceiling of the room this music made, a corridor that shot straight as an arrow and high-ceilinged as a church into some darkness out of which a groan, as if the whole house were in pain, reached out to him a hand whose fingers clutched once and then surrendered and fell back. The dimness was punctuated regularly by flashes of light that illuminated nothing. And then the animals—or what sounded like animals—began to scream. Nothing but death could have seized them and torn such screams from them.

He stood up. His son turned off the tape. "Thank you," Mr. Mortimer said. Outside the door of the studio, he stood a moment in reflection. This had been a mistake. When he had looked into his son's eyes that afternoon four years before, he had had something like a vision. He saw the boy from behind, walking straight and upright, with his hands in his pockets, sweeping his head from side to side across the path and the woods on either side of it. You could not see the end of that path, which dissolved in a green and yellow and brown dimness. Since he had entered the woods for the first time that day, he had assumed that the room or the house of his son's life included him, and he had been disturbed by the studio and the work his son did there, work that he never heard. It occurred to him that nobody else ever heard these works either, but that was no comfort at all since that only numbered him among the "nobody else." Even after he had heard the work, he still felt as if he were standing outside the studio watching his son produce "music" that would never reach him. Evidently there were woods within woods, rooms within the rooms, whose doors were always open a crack but in which he would never stand.

The business with the trains began the following week, when he and his son had gone into town on a shopping errand. As they walked past the hardware store, a movement in the window caught his eye, and he stopped to look for a moment at the model-train setup behind the glass where a twelve-car train moved slowly behind an HO-scale locomotive around an artificial landscape that included tunnels, lakes, a water tower and a depot. How pathetic, he thought at first, and then he glanced down at the top of his son's head. The boy was also looking into the window. It was hot. He looked again at the train. Inside the store a small group of boys who appeared to be around nine or ten years old were standing, watching the train move while they moved their mouths. From time to time one of them would point to something in the landscape.

"Would you like one of those?" Mr. Mortimer asked.

The boy looked into the window for a long time, and then up at his father. "Yes," he said.

When they got home and began, slowly and methodically, to assemble the train set on the table in the dining room, Mrs. Mortimer asked him if he had remembered the roast. "No," he said, and got up from the table to go back to town to the butcher shop. His son went with him. In the car, as they drove along in silence, the boy said, "Mother will not like the train to be set up in the dining room." It was one of the few times Mr. Mortimer could remember his son initiating a conversation. He was right.

"We can set it up in the attic," the boy continued.

"Attic?" They were driving past the little ballpark just outside the town limits.

"In the back of the closet in my room there is a door

that leads up to an attic room. It has no windows. It's perfectly empty."

Mr. Mortimer was surprised to hear that in the house he had been living in all these years there was a room he had known nothing about. It was another bit of information to learn, to digest. These bits of information were always "there," and from time to time they came into focus, made themselves known to you in one way or another. What made this one more interesting was the fact that it came from his son, who obviously had known about it all along and had not thought it important enough to mention. And it had *not* been, up till now, worth mentioning. What could be more empty than an empty room? But now that it existed for him it began to be charged, filled with possibilities.

It was necessary to set up some tables in the attic room, but Mr. Mortimer was glad for the room and for the trains and the landscape that he and the boy began to assemble. He was glad for the boy—even though he himself, neither as a child nor now, had ever had the slightest interest in trains or in model railroading—who gravely opened each box containing more cars or track or locomotives or transformers or some new piece of the landscape, a mountain or a tree, a billboard. The boy now began to read books on railroads and models, and as the layout expanded in size it increased also in its authenticity, its faithfulness to detail. Mr. Mortimer never, after that first question as they stood outside the hardware store, made any suggestions or interfered in any way with his son's conception of the world that was taking form in the room. He only listened, and either bought the equipment or, when his son requested, helped in its manufacture or assembly. It was a very small sacrifice to make for the happiness of his somewhat strange son.

Later, David learned from an exchange of letters between Mrs. Mortimer and her mother that the boy had never had the slightest interest in model railroading, and had begun the layout and continued to expand and develop it, almost to the day of his death, because he knew it made his father feel good to think he was contributing in this way to his son's happiness. The studio was his own world, which excluded the father even when he was in it, even when he was in the room the music made within that room, so the attic was a small recompense, a world that included everything, including even his father. So the boy continued to think up new expansions and refinements of the model-railroad world in the attic, things the father would think he found fascinating, things they could do or make or watch together, soberly requesting his father's opinion or advice, even to the wrecks.

After they had been expanding the train layout for a couple of years, the boy began staging elaborately planned collisions of the trains, at first involving only freight and tanker cars, but later including passenger trains like the silver Amtrak replica twenty cars long. He even designed and put together an engine that came apart in modules so that it could be "destroyed" in a crash and then reassembled. His father, who always watched the performance of the trains without any expression whatever, observed these crashes in the same way. Eventually, the boy stopped staging them, until François began coming up to the attic room. Mr. Mortimer had told his wife about the wrecks, which were accompanied by all the appropriate sound effects, including the screams of the victims. They disturbed him, he said, and when she relayed this to the boy he said, "I thought so. Adult people are apt to be disturbed by representations of death."

His mother looked at him.

"Lots of beautiful things," he went on, "are filled with pain and darkness. This house, next door."

"This house?" she asked.

The boy pointed to the window, where four or five bees were buzzing and colliding with the glass, killing themselves as they tried to return to the world of moving dark green on the other side. The windowsill and the floor under it were littered with the bodies of other bees.

"Death doesn't disturb you?" she asked as he got up from the table and walked to the fridge.

"No," he said, pouring himself a glass of milk.

François had been wrong, or at least partly wrong, David learned now, about the living room in the woods. It was Mr. Mortimer's son who had begun it. He had seen some young men from the college ceremoniously carrying an old couch through the woods. They told him they were wanting to give it a decent burial, and could he direct them to the refuse dump. "They were staggering around a little and their speech was slurred," he told his father as he led him through the woods to the clearing where the couch had been installed. "I told them they could bury the couch nearer than the dump, and better too, because even in death it would continue to live and serve. I told them about the clearing, but they insisted on the burial part. Then I told them we could have the burial there in the clearing. I promised to recite the ceremony—the Catholic one, since it has the most Latin. I had to teach them the responses, but it all went O.K."

Mr. Mortimer looked now at the couch. It had once been a rich crimson, and regal in size, sumptuous. It had probably come from a hotel, but then there were a number of families in this area with homes large enough for a couch this size. Most of its nap was gone and it had faded unevenly over the years, till now it was the no color of burgundy wine into which someone had poured coffee with cream. Some wealthy family had lived their life in the room with this sofa, sitting on it while the daughter's fiancé had stood asking for their permission. Sitting there, on that corner of it, while upstairs the grandmother with the skin cancer breathed furtively and that look came into her eyes as she understood and refused to understand simultaneously and her last breath gargled in her throat. From there the couch ended up somehow in the dorm room of some students at the college, who bequeathed it from class to class. It sat now against the far "wall" of the clearing, facing the path, white cotton batting spilling from its various wounds.

Mr. Mortimer stood there intrigued. Something happened to the space of the clearing just because the sofa was there. The "door" of this room was standing wide open. He could imagine an afghan draped over the back of the couch. He could imagine the students from the college coming here to sit on it with their feet up on a coffee table. They had a six-pack of beer and from time to time in their loud conversation they gestured, forgetting the beer can in that hand and spilling a little on themselves and on the couch too, which they wiped up with a couple of swipes of an open palm. He said, "It needs a coffee table."

They found a coffee table at a yard sale in Tivoli, and the easy chair and floor lamp and a heavily framed picture and an Oriental rug at the Salvation Army Thrift Store in Poughkeepsie. Placed there in the clearing, the room they made in the woods was a room in their life. Mr. Mortimer did not any longer feel the need to look into the dedications and acknowledgments in the books near his carrel in the basement of the library. He thought of keeping a large paper shopping bag in his office at the college. He would wait till Professor Hartpence stepped out of his room, leaving the door open as he always did. He would quickly and confidently move (silently) down the hall, as if he had every right to be there, and walk through the door calling, "Neal?" in a conversational tone. Then swiftly he would strip the afghan from the back of the couch, fold it neatly while his heart pounded "like a trip-hammer" in his chest and stuff it into the shopping bag which he now pulled out from under his sweater and unfolded.

But that was crazy. He had not felt a desire as passionate as that since he was courting Allison. It was to guard against just such vagaries that the structure of civilization had been erected, the city of man. That evening he asked his wife, "Have you ever thought of knitting?"

She had just sat down in front of the mirror of her dressing table and reached down to put a box of Tampax into the drawer. "Knitting what?" she asked.

Now as Mr. Mortimer walked through the woods, he could hear the sounds of an animal chewing on something hard and crunchy, like a root or a nut. The sound of it was tinny and amplified many times. The chewing was surrounded by a skittering noise, a panicky breaking or chipping of small bits of glass. These were punctuated by groans and a mumbling of Latin, and weaving through them all was a piercing, throbbing electronic jangle. As he came up to the living room, he saw his son seated on that once-red couch, the tape machine beside him. The boy touched the machine, instantly snapping the woods back into silence. "Hello, Dad, would you like to play with the train?" But Mr. Mortimer had brought a book and wanted only to sit in the easy chair and read. His son put on his headphones. They often did this now.

It was about this time that the boy began bringing books home from the library. When his father asked him about this, he said, "With most books you just read them and you're done with them. You can use the information or you can't. But with poetry it's different. It isn't like something that you use up but more like a house you live in." The house he was living in now was Emily Dickinson's.

"Why Emily Dickinson?" his father asked. They had stopped to look down into the cellar of the Barstow vault and were now walking past the tennis courts, empty in the heat of the middle of the day. A couple in tennis clothes were sitting on the grass under the shade of some trees on the far side of the courts, which lay now, idle and silent, a tall open cage laid out in red and green, the new white lines crisply defining the areas.

"She knew," the boy said, "that death is real: 'impossible to feign.' All the rules, all those lines and fences, they're all constructed. All made up—to keep death out. And it's here already, it was here all along. The way your mother is in the house all along, the one thing you know for sure is that she was there before you." He started to walk again and Mr. Mortimer followed. A couple in jogging outfits huffed past them.

"When I'm out in the living room next door," the boy said, "or upstairs in the train room or something, and I hear Mother calling me in, I don't want to go, even though I know I should. Being next door is neat, and being in the house is neat too. But saying 'No' is maybe the sweetest of all."

"Why is that?"

"I don't know. The 'No' is also an 'I know,' I guess. You know it doesn't really mean what it says. It's just a delay, a prolonging, delaying what we long for."

They went on walking for a while, and then Mr. Mortimer asked, "Would you like to try writing poetry?"

"Yes," he said after a pause. "I've *been* writing some." He looked up at his father. "Would you like to read some of it?"

Mr. Mortimer remembered the "music" he had heard down in the boy's studio. "Yes," he said finally.

When they got home, the boy went to his room and came back with the black spring binder, which he presented to his father.

"These are the LEDs," he said.

"What are LEDs?"

"Light-emitting diodes."

Mr. Mortimer did not understand why his son would enjoy using terms that the boy knew were unknown to him. When the boy left the room, his father opened the binder to the title page, which read "*Die Odes.*" Well, he thought, to the best of his knowledge the boy had never taken typing lessons. He read the first three poems as if he had been handed a text in a language he had never learned. After the third poem he closed the book and looked at the wall.

1.

Now will I die, soon. So held again to be
All so unlucky? To die, to become the night, the sheen
Of the "unlucky" shining for me alone.
The sun shines all around, meaning—?
You must not die in the night, shrunken back
To the dirt, they say. You must see in yourself the light,
Sunken in the lamp though it be, lost in the mine.
Hail, they say, afraid of the light *I* see in the world.

2.

Now I see why you warned me so, dark flames,
Spraying me with your magic blaze,
O August! O August night!
Glad sham, the volume of your blaze,
The danger of your glance, the mighty shame.

I went to the dock in the night, wild neighbor
in whom I would go swimming,

Weaving and blending myself in your quick.
That such stars presided over these rites, homely care,
The thin door that wanted all the stars stemmed
  in its own tide.

I will tell how I heard you say:
  "We might not dare to blaze out to you,
  But the dock is one of those who are able:
  The sea is not far, and here the bank is soft with ferns.
  It was here your eyes stayed in the day,
  Now you can stay like the night, white and stern."

3.

I hear the wheels turning like thunder in the river
I hear the rattle of glass, the vacuum tubes
In the gunnysack. The hawk screams in the attic of
The trees. Mother, my home is here, with you.

It was the recording and composing studio all over again, or Hartpence's afghan. This poetry was a room or a house he would see always from the outside. Worse, it was a room in his own house in which he could stand forever without being *in* it, part of the thing that made it *somebody's*—his son's, rather than just a room. Reading the poems and hearing the music was like looking into what he thought was a window but that turned out to be a mirror, revealing nothing to him—only himself, a man standing there *looking*. The boy was enfolded in the words and in the music, the rooms in which he moved took him into themselves, while the father could only stand and look.

Mr. Mortimer saw the *Die Odes*—which the boy later either included in the collection called *Love Songs to Death* or simply retitled—in April, a month before his son's twelfth birthday. In June, when his teaching duties were over for the year, he went with his wife and child to visit his mother-in-law in Harrisburg. On Sunday, reading through the paper in his grandmother's backyard, the boy saw an ad for a new model of equalizer and asked his father if he could go the next day and see if it was as good as they claimed.

"Certainly. Do you want to go by yourself or would you like some company?"

The Stereo Supermart was located in a district of warehouses and light industry, next door to a tavern. A beer truck making deliveries was double-parked outside, behind another truck. Mr. Mortimer found a parking space directly across the street, and when he and the boy had almost gotten across they heard a motor revving up behind them and turned just in time to see a black Corvette move past them with two teenage boys in it. The driver casually dropped a brown paper bag out the window, which crashed as it hit the pavement, like glass breaking. Mr. Mortimer got to the curb before he turned around again, and saw his son bending down to pick up the paper bag. As the boy straightened up, he opened the bag and pulled out a large green fragment of a Coke bottle. It was 11 a.m. He was putting the shard back into the bag when his head turned quickly to the right toward the massive back doors of the beer truck, which had just revved up its engine. The street was deserted. Mr. Mortimer was about to say something when the truck lurched into reverse, and he stood on the sidewalk, looking, seeing his son look from the truck to him with an expression that seemed then indecipherably serene, enveloped in the noises of the truck, the roaring motor and the rattling bottles. That expression did not change when the heavy bumper struck the boy on the upper thigh, just below the hip, causing his legs to buckle slightly and his body to lean and turn to the right, so that he could almost have read the advertising legend on those tall steel doors before they smashed into his cheek and nose, knocking his glasses off and snapping his head back and down, the whole of the body then, in its T-shirt and blue jeans, beginning to fall, quickly and awkwardly, to the pavement, where the double rear wheels of the truck crushed it lengthwise from the feet to the chest, where they came to a stop.

To his left something moved. It was the driver of the truck, who had opened his door and was standing now with one foot on the running board, a look of distaste and contempt on his face as he saw the body under his rear wheels. "Oh no, oh shit," he said half aloud, the words coming out slightly slurred. Then, as he moved to get back behind the wheel, he looked toward Mr. Mortimer, who was standing on the sidewalk with his hands up and open in front of his chest, as if he did not know whether to reach for the child's body or his own. His mouth was open. The driver waved at him once, a scornful downward swipe of his arm, as he finished getting back into the cab of the truck, which now lurched forward and away. Down the street.

The father leaped out between the parked cars, his right hand already sliding inside his jacket to the shirt pocket and pulling out the ballpoint pen and clicking the button, his head leaning over the parked car in front of him, craning to see the license number of the truck and mumbling it to himself as he searched his pockets for some scrap of paper—repeating it meanwhile, and finally writing it in big block letters on the palm of his left hand. The truck swerved wildly around the next corner, and he was left alone there on the street. He did not want to look down. He had seen enough to know that nothing could be done. The boy was beyond help, beyond pain, beyond any call.

When the police came, they were very good, very professionally courteous and understanding. One officer leaned into the back seat of the police car in which Mr. Mortimer was glad to be sitting, since he didn't feel any strength left in his legs, and asked him if he would like a drink or a glass of water or something. "Yes, thank you," Mr. Mortimer told him, though he refused the tumbler of whiskey someone had brought him from the tavern. The inside of the police car smelled of Naugahyde and of the oil from the riot gun whose barrel he could see rising up into the frame of the windshield. This is how criminals must feel, he thought, sitting in the back seat of a squad car under the power of the law. It was an odd thing to think about, he thought, as he drank, feeling the water fill his mouth and slide down his throat.

At the police station, where they asked him more questions and had him complete some forms and sign others, he felt even more strongly the machinery, the working structure of the law, this whole building, all these computers and teletype machines, all these people, working for him, to enforce the law. One heavyset man in a tan suit told him, "If everything you're telling us is true, it sounds like at least felony hit-and-run, and most likely manslaughter."

"Of course it's true," he said quietly.

"Do you have any witnesses?"

Behind him, all around him, he heard the hum of the air conditioner, and over that the bustle of voices and teletype machines, the generalized background noise of an office where people worked.

The hard part, he thought, would be telling his wife and his mother-in-law. But when he got out of the plainclothes policeman's sedan in front of the house the two women were waiting for him, pulling aside the curtain in the living-room window and then appearing in the doorway. Their faces were drained and empty, they could not believe it, the police had called them but they had not given them any of the details—"Charles, a *beer truck?*" It was then that he realized he would begin to cry. He did not know how to cry, and he felt his face contort itself, making him look, he thought, grotesque, and making him feel doubly foolish: he had been prepared to comfort the women. His son. He was twelve years old. None of this was true.

He was thankful, with that part of his mind that still functioned clearly, that there were practical matters to be taken care of. His car was still parked across the street from that stereo place; he had to do something about a funeral. They would ask him, he thought, to select clothes in which his son's body would be dressed. Someone would put those clothes on his son's body. How? The arms and legs of the body would be stiff. There would also be an inquest of some kind or a hearing, and there he would be presented with the opportunity to testify, to tell how it happened. And then that man, that driver of the beer truck would learn, would learn the meaning of the law. He had *hulked* on the running board of the truck like an animal, his eyes glassy, his speech slurred.

The next morning Lieutenant Roventini, the heavyset plainclothes officer who had spoken with him yesterday at the police station, called to give him a progress report. Through the license number they had located the truck, which was registered to a local beer-and-soft-drink distributor. The people there had cooperated fully and had given him the name of the man who had been driving the truck yesterday on that route, "A Mr. Anthony Blaquere, been with the firm thirteen years. But, Mr. Mortimer, he says he didn't run into anybody."

"But—"

At the preliminary hearing Mr. Mortimer told what had happened. The hearing room was much smaller than any courtroom he had seen in the movies or on TV. It was one of what seemed like half a dozen rooms on this floor where hearings of one kind or another were being held. People he did not know were sitting in clumps around the spectators' area listening to the proceedings of some other case. A middle-aged woman worked the little keyboard of her shorthand machine while she chewed gum. At first this offended him, but this was precisely the impersonality on which or of which the structure of the law was built. The fake walnut paneling bothered him because it seemed unnecessary. It was not only an unsuccessful attempt to suggest the austere splendor of the law but a halfhearted one, like a poor set in an amateur theatrical production.

Mr. Mortimer now saw that Anthony Blaquere was bigger than he had remembered. As he walked forward toward the stand the muscles of his shoulders rolled under his coat, and when he sat down he took up the whole space of the chair, the bulk of his body rising up thickly to his thick neck and his close-cropped head. His small, dark eyes looked out of his face with a kind of brutish repose. Mr. Mortimer realized that the man did not feel in any way threatened by any of this and looked out at his accuser with the scorn of obliviousness.

"I made my deliveries to Mortenson's. When I got back to my truck I finished up the paperwork. There was another delivery van parked in front of me, so I threw it into reverse and pulled out. The next morning the police call me and tell me I ran over some kid. I'm sorry the kid was run over—who would want to see a kid run down? But he must have jumped out behind the truck—even with the reverse beeper going and everything."

"Didn't you feel the impact?"

"Have you seen that section of Harrison Street?"

"Why?"

"It's nothing but potholes and debris in there. I don't think I even thought about it, but if I did I must have figured I had run over some debris."

"Mr. Mortimer claims you were drunk, that you stopped your truck and got out of the cab and looked back, that you—"

"That's a lie! That's a lie! I'm really sorry about his kid and everything, but he's calling me a murderer and, hey, that doesn't go. That's a lie."

From his seat twenty-five feet away Mr. Mortimer could feel the hatred billowing out to him from the man, an almost palpable presence. Now the man was actually looking directly at him, punctuating his outraged sentences with short thrusts of his finger. His indignation was so compelling that for a moment Mr. Mortimer actually found himself convinced. Then he felt his jaw drop open, and a moment later felt something surge through his whole body, almost as if his chair had been hit from behind. He clenched his fists in his lap. This is what hate feels like, he said to himself. This bullet-headed man with the sneering face and the jabbing finger had killed his son and was now, in a public courtroom, in full view of the world, contemptuously, barefacedly lying and calling *him* a liar, challenging him. This could not be happening. None of this was true.

He still could not believe it when he heard the announcement that there was not sufficient evidence to indict the defendant.

As Blaquere came up the aisle from the stand, he was joined by two men in suits. Together they seemed to clog the space of the aisle. The boy's father looked into the driver's eyes, who looked back at him with what he took to be cold disdain. Mr. Mortimer was afraid that the man would hit him, here in the courtroom full of people. He was not afraid of being hurt, but he was afraid of being shamed further, that he would try to strike the man back, and that his ineffectuality would once again be displayed to everybody, like a man kicking a building and breaking his toe. He continued to clench his fists, and slowly realized by the slight ache how tightly he was grinding his teeth and what sort of grimace was contorting his face.

The hallway outside was crowded with people milling around, and as Mr. Mortimer made his way to the elevator he did not see the man until he felt his right arm grabbed and held above the elbow and then felt his left shoulder pushed against the wall. Blaquere was glowering down into his face and snarling, "Professor! I ought to cut your liver out, you lying motherfucker!"

The other two men were trying to pull him away, trying to get some leverage to move him by his arms and saying, "Tony! Get the hell away from him! Tony!"

As the man let himself be moved away, he was still shaking his finger in Mr. Mortimer's face, growling, "You almost killed my poor mother! She just about took a heart attack!" The other two men pushed the truck driver along.

As the doors of the elevator closed behind them, Mr. Mortimer continued to lean against the wall of this corridor in the Hall of Justice. The whole thing had been remarkable for the quietness of its violence. Only a dozen or so people milling around them had noticed it, and even their surprised immobility was now already blending back into the general stir and movement of the hall. A large black woman looked at him and then at the door of the elevator, shaking her head. "Animals!" she said at last. When he tried to walk, he found the muscles in his thighs and calves were too fluttery to support his weight, and he had to stand where he was, leaning against the wall, for a few minutes. He realized that he would have to tell all *this* to Allison and her mother too.

The suggestion to see Callahan, the family lawyer there in Harrisburg, came from Mrs. Baxter, Allison's mother. Mr. Mortimer went that afternoon and told him the story. At first the lawyer was puzzled that no other witnesses were called. "Didn't anyone question the tavern owner or the people in there?" he asked Mr. Mortimer.

"No, why?"

"Well, you say he had been drinking. Maybe there was somebody in there who saw him or spoke with him, someone who might be willing to testify to that effect. But didn't the coroner or deputy coroner present a report?"

"The coroner's report was made a part of the record."

"This is all extremely irregular. Even an insurance company would have made a more thorough investigation than the one you've described. I don't quite know what to tell you. Certainly you could file a civil action for wrongful death, naming both the driver and the owner of the vehicle. I doubt if the driver has much, but if the company were held liable they obviously have assets.

"This is not a neat case, Mr. Mortimer. You see, the preliminary hearing tends to exonerate the driver, and in order to press the case we would have to somehow impeach or reopen those proceedings, and that means fighting City Hall. I don't know what to tell you. This is all most—messy. We would have to bring in our own private investigators. Messy and expensive."

The phone rang. While the lawyer spoke affably into it, Mr. Mortimer imagined himself talking to the bartender in Mortenson's Tavern. The flat, cool dimness was a different world from the hot street outside. Over the mirror behind the bar a lighted glass diorama showed a miniature beer wagon pulled by a team of tiny Clydesdales. No one was there except for the bartender, perched on a stool at the far end, eating a liverwurst sandwich.

"You know Tony Blaquere?"

"What's it to you?"

It was absurd. He looked out the window at the trees of the park across the street, all reaching their branches out into the brightness of the air, as if to contain it, which they did, some of it, moving intermittently among their green darknesses. Now the man across the desk was saying, "Good talking to you, Jack," and hanging up, and turning to give Mr. Mortimer a look of calculation.

"There is something else about this case that bothers me," he said. "An investigation like this is routine, it *should* be routine. Policemen are human, they can be distractible or careless at times, even incompetent at times. But this sounds too—well, the standard procedures don't seem to have been followed, and one has to ask why. Why weren't they followed? The pattern suggests—well, it suggests that somewhere along the line a decision was made—by somebody or some bo*dies*—to *not* follow the standard procedures. Do you see what I'm driving at?"

Mr. Mortimer looked at him across the desk. The lawyer was leaning back in his chair, turned at an angle, yet his head and shoulders leaned forward, and under his bushy eyebrows his dark eyes were hard and cold. Mr. Mortimer felt he was being taken much too quickly over ground he had never even known was there. He was always stumbling onto something new just when he thought he was about to catch up. He took a breath and said, "Someone is covering up?"

There was a pause.

"We don't like to make accusations. But certainly if the—how can I put it?—the *incompleteness* of the investigation has elements of deliberateness, then, well, we are going to encounter a good deal of resistance."

Mr. Mortimer looked at him. He saw again the man on the running board of the truck, waving him scornfully away. He saw again the two men in suits hustling Blaquere—his animal bulk—into the elevator.

"No organization is going to welcome an exposé of their incompetence," the lawyer said. "Think how much more resistance there is going to be if there was deliberate malfeasance. This may prove to be, as I've said, very expensive, and very difficult. Mrs. Baxter has been a friend and a client for many years, and we would represent you, if you chose to go forward, but I have to tell you now, I am not *eager* to join this battle."

Mr. Mortimer continued to look into those eyes, which now seemed to him tired. "In ancient Rome," he said, "they would have hanged the driver and thrown the truck into the Tiber."

Callahan's bushy eyebrows went up. "Happily, we're not in Rome."

"If you did take the case, would you pursue it thoroughly, wherever it might lead?"

"Mr. Mortimer, we would never take *any* case without that commitment."

"What forms do I have to sign?"

Callahan's manner changed instantly. He leaned forward with a smile, saying, "Nothing for now. I'll have my secretary draw up a letter. We'll need a retainer—five thousand should do for now. You can give me a postdated check if you like."

When they got home, Mr. and Mrs. Mortimer found that Emily Dickinson had been right when she spoke of the time after death—for the survivors—as an "awful leisure." Mr. Mortimer, who did not play tennis as much as his wife, and who did not keep the women's chorus running smoothly, found himself shrinking from reminders of his son at the same time that he sought out and clung to reminders of his son. Finally he made the decision and either destroyed or disposed of everything. The clothes he gave to the Salvation Army, the records he donated to the college library, the tapes he stored in the attic train room, which he never again entered. He never again went into the woods next door or into the studio in the cellar.

He started getting reports from Callahan. Often these

were several pages of single-spaced typing, together with photocopies of relevant documents. Anthony Blaquere was thirty-seven years old and had never been married. He lived with his mother in a working-class neighborhood, in the bottom half of a duplex, which she owned. The top half was rented out. He had been arrested twice in the last five years, once for assault and battery—a fight in a bar—but the charges were dropped when the other man withdrew his complaint. The other arrest was for drunk driving, for which he was convicted, given a suspended sentence and assigned to the police department's Driver Training Program, which he completed successfully. He did not go out much, except for an occasional movie, and even more occasionally to a porno movie. He frequented three bars in his own neighborhood. He had an older brother and an older sister. The sister was married to a Marvin Jacobs, who managed the distributing firm for which Blaquere drove a truck.

The garage that did the routine maintenance and repair on all the distributor's trucks revealed that they had repaired Blaquere's truck the day after the accident. Blaquere himself had brought it in, even though it still had two thousand miles to go before its next scheduled tune-up. He claimed it sounded like it was missing on one cylinder. Nothing was really wrong; somehow one of the spark-plug cables had pulled loose. "Oh, while you're at it, you want to take a look at that reverse beeper? It went out on me last week." The beeper was repaired. Photocopies of the deposition, the repair order and the invoice were enclosed.

Mortenson's Tavern, next door to the Stereo Supermart, was the fourth bar on Blaquere's delivery route for Mondays. Thomas Mortenson, who owned the place and worked the day shift, revealed that Blaquere usually made his deliveries in fairly prompt fashion, but that fairly regularly—

once every six weeks or two months or so—he would come in obviously hung over. On those days he would have a boilermaker or two. Hair of the dog. On the day in question he had complained of a terrible hangover and had his two boilermakers. "He is a big man," Mortenson said. "He can hold his liquor pretty good." No, he did not know if the driver had been drinking before he arrived that morning—how would he know?—but yes, his speech had been a little slurred.

The man who worked the morning shift in the Astor Club, the first bar on that delivery route, said that Blaquere had simply wheeled in his hand truck, picked up his empties and left, almost without saying a word. But at the Tahiti Lounge he had had two shots of whiskey and a beer, and at The Spot he'd had another boilermaker. "Getting an early start, hey, Tony?" Blaquere held his close-cropped head in both his huge hands. "Got to kill the demon," he said. "It feels like there's six dogs in there fighting it out. Only way to do the beasts in," he said, holding up the shot glass. Photocopies of the three depositions were enclosed.

In the autumn Mr. Mortimer taught his classes as usual, except that it was pretty obvious to everyone that he was distracted most of the time. A few students complained that he would trail off in the middle of a sentence sometimes in class, and end up staring out the window. At other times he would lash out sarcastically "for no reason at all" at some student "who hadn't done anything to deserve it." The dean and the president called him in and suggested he take a leave of absence.

"I had thought of doing that," he said calmly, "but this is what I need right now."

It was the first time that either man had ever known him to admit that he needed anything. He went on teaching his classes one way or another, and spent even more time in his carrel in the library, but found he could only read articles in journals. He felt his mind to be in a kind of neutral, numb. He could ingest ideas and discriminate between them as sharply as ever, but he could not produce new ones. Now again he walked the aisles of the library, randomly taking books down and reading their dedications and acknowledgment pages, a wistful feeling passing over him whenever they mentioned the author's children. "To my sons Timmy and Robin," "To my daughter Kathleen" and "Some acknowledgment is due also to my children, Robert and Alicia, who put up with all that typing at all those late hours."

He found that he and Allison were being invited to dinner more often than usual, and that all the people they knew were being terribly bright and cheerful. At these dinners there would as often as not be a child, either at the table or upstairs in bed, and everywhere throughout these houses were the evidences of an ongoing life that included these children—cribs and high chairs, bicycles, wagons, doll houses, musical instruments, a little toy seal he found in the bathroom, made out of some tinny kind of metal. When he rolled it along the windowsill, the striped ball on its nose turned. It would have been so easy to put it in his pocket. It couldn't be worth more than a dollar; it would probably not even be missed. He let the wheels spin themselves out and put it back on the windowsill.

He found that he could not any longer think about his son without conjuring up also an image of Blaquere, the man hulking like a bear or a gorilla there on his chair at the hearing, there on the running board of his beer truck. And with these images came a feeling of cold tenseness in his skin and his limbs, his teeth clenched, his cheeks tingling, his arms tight, his heart beating slow and hard. When he saw Blaquere on the running board of his truck, he saw also, out of the corner of his mind, the blur of blue jeans and T-shirt and the spreading redness on the pavement.

There was more news from Callahan. Marvin Jacobs, Blaquere's brother-in-law, had a brother Michael, a captain in the Harrisburg Police Department. Blaquere's older brother, Ryan, was married and lived in the suburbs with his wife and three children. He was a sergeant in the same police department. "I warned you at the outset," Callahan wrote, "that we would very likely encounter resistance, and we are. Every petition we file takes three times as long as usual to process, and is often rejected on the flimsiest technicality or no technicality at all. These all have to be filed again and the process begun again from scratch. I don't think I can get a hearing set before the spring, but by then we should have as strong a case as could be made, I think, and should be able to move on both fronts, the reopened criminal case and the civil suit for wrongful death. Enclosed please find a photocopy of Ms. Krantz's deposition."

Rachel Krantz had worked at the Stereo Supermart on Harrison Street for three years. She was now twenty-seven years old and had recently broken up with her boyfriend, who worked in the accounting office at Blue Cross. Because of the breakup she had decided to move back in with her mother in Philadelphia, and maybe finish up her degree in social welfare at the University of Pennsylvania. On the morning in question she had opened up the store as usual at ten o'clock. She liked the mornings best because almost no customers came in till around lunchtime, and she could leave all the various display systems turned off. After noon there would be three or four stereos going full blast, all tuned to different stations.

"I was sitting in the front, going over the week before's receipts. I remember because it was my last day there and I realized I was actually feeling sentimental about the place.

The other two people I worked with were in the back. I heard a loud roar and looked out the window. It was one of those trucks double-parked outside. The beer truck went into reverse. It sort of *lurched* backwards and then it stopped. The driver opened the door and got out or stuck his head out and looked back. Then he waved at this other man who was standing on the sidewalk, and then he got back in his truck and drove away. Just then the one customer who was in the store came up and asked me about speakers and I went back to the listening room with him. We must have been in there, I don't know—forty-five minutes, an hour? But he did buy a set of AR 14s. I didn't even hear about the accident until now."

The hearing was set for May. Mr. Mortimer drove to Harrisburg in a rented Mercury Cougar. He stayed with Mrs. Baxter, and on Sunday morning he and his mother-in-law drove out to the cemetery, which they entered through a neo-classical arch in a vaguely Roman style. Mr. Mortimer had never been bothered by cemeteries, and as he drove between these gently rolling green slopes the neat rows of graves looked to him as if they were flowing over the hills like waves. Their orderliness made him think of one of those waves his father would have studied, all of whose parts were in some sort of harmonic relation to each other. Was that it?

But even as he was pleased by the mathematical neatness around him he was struck by the anonymity of the markers that in fact marked or distinguished nothing. He could read their names, he could read their dates, they would still remain anonymous. He would never stand in their kitchens chewing on a cold ear of corn from the fridge at eleven-thirty at night after watching the late news. Other people's lives. Other people's deaths. In a broad, shallow bowl in these hills, like a wide meadow that might have been a wheatfield, the wind making a sea surface of its hair, in one of the graves in this row of graves, his son was buried. As they started to leave, approaching that Roman arch along the wide, boulevard-like main road, he remembered the one trip he and Allison had made to Paris, walking up the Champs-Elysées.

After lunch he spoke with Callahan on the phone and arranged to meet him outside the courtroom. He and Mrs. Baxter had an early dinner. He watched some TV for a while. He got up and said he felt restless and thought he would just drive around for a while. He looked up Blaquere's number and address in the phone book in a gas station. He drove there in the gathering evening, and cruised past the house once. Then he went around the block and parked across the street, two doors up from the house. The duplex was like a twin of itself, one sitting on top of the other. Identical screened-in porches top and bottom. In the bottom porch two figures were sitting in lawn chairs. They must have been talking, because from time to time one of them would gesture. The air was muggy. The house was covered with asbestos siding that was brown but streaked or mottled—to resemble what? Soon Mrs. Blaquere, whom the neighbors called Mrs. Blacky, would get up and go to bed. Anthony Blaquere would go to the kitchen and take a beer out of the fridge. He would put the flip-top ring in the ashtray and take his first drink there in the kitchen. TV was a pleasure he indulged himself in, and this forestalling of the pleasure was somehow good too, real good, an assertion of willpower. He carried the can into the living room and turned on the set, its jumpy blue glow filling the dim room. Mr. Mortimer drove away.

In the morning he and Mrs. Baxter made themselves a good breakfast, and then he drove downtown. As he got out of his rented car in the parking lot of the Hall of Justice, he looked up to see Blaquere coming with the same two men toward him and the building behind him. He walked out from between the parked cars and saw that the other man had now spotted him, pouncing his snout-like nose forward and snarling, pointing and jabbing his finger the way he had done in the hearing room. He was trying to speak but his fury made it impossible, and the choking stammer made him even more enraged. He waved again, one downward, scornful swipe.

Mr. Mortimer pulled the .45 from his belt and snapped off the safety with his thumb. He squared off and dropped down into a crouch, the way he'd seen the TV cops do it, holding the big gun with both hands, his knees and elbows bent to take the recoil, aiming at Blaquere's chest. Out of the corner of his eye he saw Callahan running toward him waving and calling, "No! No, Charles, no!" The kick of the recoil did not surprise him, but the noise of the blast did; it jolted him and made him blink. The noise was a room—formal, even elegant, but absolutely bare. The large window on his right, with many panes, flooded the room with light—its white walls, its hardwood floors, the creamy richness of its baseboards and moldings. The room was long, and on his left, at the far end, an archway opened onto another room and gave some suggestion of the size and complexity of this structure. A foot or so behind him was one of the end walls of the room, and at the far end of the room was Blaquere. The impact of the bullet had knocked him back against a car. His eyes were still furious but now also puzzled as he lurched forward. Mr. Mortimer fired again, keeping his eyes open this time. The man looked as if he'd been kicked in the chest. He sat down on the pavement with his back leaning against a car, his legs twisted under him, and then that look of panic came into his eyes when he realized that this was, for him, what all his life he had called "dying," a word that up till that moment had had no meaning.

• • •

The letter from Michael Harrison came in the morning on Saturday. Later that evening they were going to Anselm's goodbye party for them, and on Monday they would head back to California. *Les Champs Magnétiques* was "finished" except for some minor editing changes. Anselm had already begun the process of setting up a traveling installation-performance schedule that would, ideally, include Boston, New York and Washington-Baltimore. The piece would live or be housed in various lofts or galleries or dance studios in those cities, if everything went the way it should. David would take care of the West Coast and do what he could with places like Dallas and Boulder and Santa Fe. As he opened the mail, the party tonight seemed both anticlimactic and premature. The piece *wasn't* in fact finished. Of course it would never be finished; that was part of what it was all about—everything can be packaged, nothing will fit. Yet there was something satisfying about the feeling of closure, of a neat ending, that he gave up only reluctantly.

Michael's letter was cheerful and chatty. His recitals in San Francisco had gone very well, considering the usual semi-pro circumstances—pianos with slow or muddy action, squeaky pedals, halls with poor acoustics, and so on. He had played some of David's songs and had seen all their mutual friends, including

3 / DANIEL

the Riordans.

"You probably already know of Annie and Daniel's breakup, but if you don't I'm sorry to be the first to let you know, since the information should have come from one of them. The whole business was quite sudden for everyone, but it goes something like this: Daniel fell in love with an old student, Connie, with whom he had been having an affair for the last couple of years. He announced to Annie only in June that he was leaving her and moving in with Connie. Annie was naturally thunderstruck by the news. I think she is just now coming out of the shock of it all. She's keeping the house, at least for now, and Daniel has moved into Connie's apartment in Presidio Heights. Both seemed pretty well, considering. Annie was alternately up and down when I visited her: excited about the prospect of independence and yet feels understandably sad about the split. Daniel was a bit subdued, but the two evenings I spent with him were uncannily *normal*, given the circumstances. He has been staying away from any socializing, although he and Connie and I went to Chris Garfield's concert last week, which, I gather, is the first time he had been 'in public' with her.

"I don't know what to say about all this—it's really not easy to assimilate the split of two friends who, despite obvious irritations, had developed a kind of special language together and who looked more solid year by year. I don't know Connie well enough to even say anything about her. She is younger than Annie (I think she's about twenty-three) and amazingly beautiful, a classic Swedish blonde, a model—though she must have some brains, since she was being scholarshipped through school. She was very silent during my visits with them, which is understandable, I suppose, since she doesn't yet know *any* of Daniel's friends and is essentially the 'new kid on the block,' very conscious of not wanting to make a bad impression. It will take some time to tell what she's about.

"I'm very sad about Annie—"

When Danny came down, late as usual, to breakfast, he took one look at them and asked, "Hey, what's the matter?"

"Oh, Danny," Jane said, hugging him to herself.

David watched his son move ambivalently into the hug, repeating, "Hey, what's up?"

"We just got a letter from Michael in San Diego. Daniel and Annie are splitting up."

He watched Danny's eyes go wide in astonishment. He looked at the table where the two yellow sheets of Michael's typed letter lay, unfolded now. They had been invisible to the boy at first, David thought, just part of the general litter that covered the table and seen out of the corner of the eye as part of the breakfast mess. Now they were, sprawled between the sugar bowl and the milk carton, the focal center around which the rest of the table and the people around it, the bricks of the wall, the window, even the bees buzzing and knocking themselves dead against the glass, organized themselves.

"Oh no," Danny said.

They had named him after Daniel. The Riordans were

his godparents. Twenty years. They did not have any closer friends. They had bought the cabin on the Russian River together—"time sharing"—though they almost never used it except as a foursome. It was their "country place," where they would go to be completely, comfortably wacky together—get pleasantly drunk, smoke dope, walk in the woods. They went skinny-dipping together in what they took to be a secluded bend of the river—till two guys floated past in a canoe rigged up with a sail, an ice chest and even a little Coleman stove, selling beer, soft drinks, ice cream and hot dogs. Their prices were printed on their sail and they weren't wearing any clothes except for their plastic strap-on bow ties. Daniel would bring his stories to the cabin, the ones that were later published as *The External World*, and they would listen to him read them out on the deck in the summer or around the Franklin stove in the winter. David would bring his guitar and they would listen to something new—if he had it—or they would all sing whatever they were in the mood for.

They had concocted alter egos for themselves up at the cabin. Daniel and Annie became Dr. Zarkov and his protégée, Dale Evans, who could never quite understand why the doctor always wanted her to take "naps" with him in the afternoon. "We never sleep," she said. Or why he wanted her to do all the housework—all the dusting and moving around of vases of flowers and such—while wearing nothing but a little white lace apron like a French maid. "I don't see how he expects me to get *any* work done." David and Jane were Chuck and Juanita Roast, from the suburbs, the Redwood City Roasts. Juanita (née Goldenberg) was a nisei schoolteacher from Stockton who ran a tight class and didn't take any "attitude" from her junior high school students. But out here in the country she could let her hair down and put her feet up and be just plain folks. Chuck (a mean man with a barbecue skewer) had played football in high school (Chuck the Truck) and now sold cars: "This baby's got factory air, four new tires and a rebuilt transmission. You can not go wrong. I'll *give* you half of my commission. You want a car, I want a sale: am I going to let my first sale of the day slip away for two hundred and fifty dollars? *Hell* no. Never mind that sticker price: you just made yourself two and a half."

They had all slept on each other's couches as their marriages had gone through their stresses and strains. He remembered coming in with Jane one night at midnight from a concert and finding Daniel sitting on the couch, his feet propped up on the coffee table, a pile of exam blue books in his lap. The top one was open and he was going over it with a red pen. He had looked up at them and said, "Life goes on." Danny had let him in and given him the blankets that were piled at the end of the couch. David remembered sitting in the kitchen of Daniel and Annie's house with a drink in his hand, fighting back his tears and saying, "I don't know what she wants. Do you know what she wants?"

So, selfishly, he was sorry because all that was gone. It depressed him to think he would never have that special kind of fun and that special sense of being together in which things did not need to be explained. And it depressed him also to think of Annie, who was not a young blonde model who was amazingly beautiful. She was a handsome forty-year-old woman, curator of the Oral History Archive of the California Foundation, whose talk was filled with stories of the Gold Rush days or early farming in the Sacramento Valley, the conflicts with the railroad barons or fishing and shipping in the Delta. David loved to listen to her tell these stories and once asked if he could hear the original tape of an oral-history interview. On the tape her questions had been insightful and sensitive, tactful. The old man she was talking with was a thundering bore. And now she would be a forty-year-old divorcée, and it pained him to think of her being bright and vivacious on a dreary parade of dates with creepy middle-aged men who did not appreciate her intelligence or her warmth or her wit.

Or her strength. It was clear to both him and Jane that she had been the real center of gravity in that marriage, that she could let Daniel be Dr. Zarkov because she knew damn well that little Dale was actually holding up the structure. It was she who managed the finances and made sure the bills were paid on time, and the taxes. She had gone through a scare five years ago when she found a lump in her left breast, and she had gotten through it just fine, with a kind of respectful cheerfulness. "Sure it can kill me, but it can't bring me down."

And Daniel. He felt sad about Daniel, and angry too. He had to keep reminding himself, He does have a right to try to find happiness. But with a twenty-three-year-old model? How dumb could he be? He was going to put Annie through all this shit for some bubble-headed twit who could make him feel like a kid again, that he was not really forty-five and starting to go gray. "In what other profession," Daniel had said, "can you watch yourself grow old so relentlessly? Your students are always the same age, always in their early twenties. Meanwhile you—are—fading out." Daniel had kept himself in shape, David had to admit that: he ran thirty miles a week—*every* week—more than David could lay claim to. But obviously this was some sort of mid-life crisis that he should have been able to get through without causing this much turmoil, this much pain, this much irreparable damage. Did he really expect this thing with Connie to last after they changed it from an affair to a live-in, day-in-day-out reality?

And then there was the "two years" business. "Daniel fell in love with an old student, Connie, with whom he'd been having an affair *for the last couple of years.*" David had had his affairs and figured Daniel had too. For that matter, he figured Jane had probably had her share: at thirty-seven she was still very attractive, and even though she was heavier now she was actually, he thought, sexier than ever. But his affairs had all been casual, almost all of them one-night stands, matters of opportunity. Like the time after the chamber concert in Montreal. At the conservatory in San Francisco he'd had dozens of opportunities, but he had made a sort of vow never to get involved with his students: it was not either professional or ethical, and he was a little relieved that Michael had said an "old" student (obviously he meant "former"). That said *something* for Daniel's discretion.

But two years! How many times had David seen him in the past two years? How often had they spent whole evenings together, whole days and weekends at the cabin? And not a word, not even a gesture to indicate this other thing with Connie, this other life. And it had been a complete shock to Annie, so for two years Daniel had in fact been leading two lives, and one of them had been kept somehow absolutely secret, excluding everybody but himself and this Connie person. The friendship, David had thought, was itself like a marriage or a family, it was an ongoing music that created a room in which everything revealed itself in a comfortable light. And now all these surfaces, that he had assumed were so *open*, that gave such familiar evidence of their inner being, turned out to be opaque.

And Daniel. Did he have his being for these two years in his marriage with Annie and in his friendships, so that this secret affair was a sort of supplement? Or did he have his being in his affair with Connie so that *that* became the frame within which he lived, whose details—the color of the rug, the pattern of the wallpaper, the house plants in the corner—helped to make him up, so that his marriage and his relationships with friends became a pretense, a fiction, a role that had to be sustained at the cost of constant awareness and vigilance, the details of behavior—the drumming of fingers, the scratching of the head, the silences and the idle chatter—becoming conscious choices, the way an actor will choose to build up a character out of slouches and shrugs. Where was it that Daniel had his being?

On Monday they got off to a late start, and at nightfall they were still trying to make up for lost time. They eventually stopped in Brookville, Pennsylvania, at a motel that looked across a large park to a minor-league baseball stadium. From this distance the noise of the crowd sounded like the humming of bees. David thought about Daniel and Annie.

Daniel was a popular teacher. He taught seminars in fiction writing and he also taught large lecture courses on The Modern Novel that drew 150 to 200 students. Two and a half years ago he had given a course on Realism in the Modern Novel that drew 250 students. The class had to be moved to a larger lecture hall. Daniel had loved it. "I'm a ham," he said. And it was the fact of his being the center of all this attention, a star, that excited Connie and attracted her to him. That and the combination of his hard, athletic body and his graying hair.

She was tall and willowy, her broad Swedish face framed by her ash-blonde hair, which she wore short. The first time he saw her she was handing him one of those computer cards the students turned in to register for a class. "Hi!" she said brightly. "I'm Connie Olson."

He had been putting his books and notes away after the third lecture of the course and the hall had almost emptied of students. Automatically he took the card she was handing him and looked at it. Across the top the computer had printed her name.

"Hi," he said, smiling, "but you should have given this to the T.A."

"Oh, I've already done that. I just wanted to introduce myself. I've just transferred here from J.C. and I thought it would be a good idea to get acquainted with my teachers, especially in these big lecture classes." She looked him straight in the eye with an easy, comfortable directness. Her eyes were dark blue. "Your class was recommended to me by a woman in the Advising Office," she said, "and I want to go back there today and thank her for a good tip. You're *good*."

"Thank you," he said modestly.

"You don't have to be modest. The people over in Advising know you're good, your students here know it. You *are*." She was smiling warmly.

She usually wore blue jeans and a T-shirt or a turtleneck. She did have one hell of a body—long and lean, with a firm, good-sized bosom—but she did not flaunt it. A couple of times she came to class dressed up, but then she looked like a smartly turned out secretary going to work on Montgomery Street—a plain black skirt, a tan blouse open at the throat and "sensible" shoes. She often stayed after the lectures to talk to him, and she always smiled as she came in and left, but she did not make a pest of herself. She was not a groupie.

He wondered when she would start coming to see him in his office and how he would tell her that he could not put any move on her while she was enrolled as his student. However, if she could wait till the term was over . . .

She did not come to his office until just before the term paper was due, in the last three weeks of the course, and then her questions were all legitimate—"A Character Analysis of Theodore Dreiser's Carrie Meeber," by ("Yours Truly,") Connie Olson. It was a B-plus paper. She would always write B papers, and her male professors would always give her a B-plus. He thought that if she really got herself organized and focused her attention better she could probably earn the B-plus flat out. But that would be it. She had a B-plus mind. But now he was being condescending, and trying to distance himself from his stiffening prick. And it wasn't that he *just* wanted to fuck her. There was a tough, energetic thereness about her manner that he enjoyed, a directness that was comfortable, without being either aggressive or needy.

She never came to his regular conference hours, when his office was jammed with students, but always knocked on his door in the afternoons, when he was alone. With anyone but a pretty girl he would have told the person to come back later. Over the years he'd had a half-dozen or so affairs, mostly with former students, though once with a visiting professor of Italian, a dark-eyed brunette from Torino who quoted Dante to him while they fucked. These were all safe, innocent affairs with women who were clearly passing through, up-front on both sides as *temporary*, even though they were, he thought, genuinely affectionate. He never broke his rule about current students, and he never let any of it hurt Annie, who never had a clue.

He did not know if he loved his wife anymore. He enjoyed her and appreciated how she took care of the house and him. She was comfortable as an old shoe, as their old house, whose details all took care of themselves. He supposed he was being selfish, but none of this had ever hurt her. It had not even cost her any of his own affection or sexual energy, since after sleeping with these young women he felt himself not exhausted but refreshed. Some of his best lovemaking with his wife had come in the evening or night of the same day he had made love to Jackie or Karen or that redheaded Helen, who came three times pumping herself up and down on his lap as he sat behind his desk in his office.

Now Connie Olson sat in his office as the afternoon light held the cube of the room in a sort of suspension, drenching the bookshelves that surrounded her and him, and told him about herself. She too was a native of San Francisco, and had graduated from Galileo. She was the youngest of three children. She lived with her mother in an apartment in the Marina. "I mean, I live in her apartment *building*. We have separate apartments. See, she is the manager." Her father had been a C.P.A. who died of a heart attack when she was fourteen. "He had a drinking problem. He would come in to kiss us good *night*?"—the hitch in her voice turning the statement into a question—"and I could smell the liquor on his breath. Once he leaned over like *that?* and he actually fell on top of me." The mother never remarried.

She had put in a year of business school and gotten a job as a secretary in the International Office of Bank of America, down on Montgomery Street. "I looked around and I saw the people around me—I mean, not the other secretaries or anything, but the higher-*ups?* They all had something they were coming from, they were always referring to, like a whole other world. They were cultured, they were educated, they knew what they were talking about. I mean, these were all the people over me. And I thought, I'm going to get some of that, and here I am." She was on some sort of educational leave of absence from B. of A., and had discovered a scholarship for women of Swedish-American ancestry. She loved literature, especially fiction, but she was taking mostly Business Administration courses, so that she could go back "and get promoted. I know I'm going to make it."

Daniel sat back in his chair, trying to keep from looking directly at her body, and thinking how could he get this girl to come back and see him after the term was over. Underneath that blonde all-Swedish-American-girl-next-door exterior he could sense a heavy grace in her movements and in her stillnesses that went with a rhythm like "feline leisure of lynxes."

When the term was over, he waited for her to come back and visit him. He would tell her, "Sure you're beautiful, but there were probably three or four women in that class more beautiful than you [it was not true], but you've got something that none of them has, some kind of liveliness, some kind of animation [this was true]." He would reach out and take her hand, and after a long, tender look she would lean forward reaching out her mouth to kiss him.

But she never did come back to visit him in his office, as she said she would do the last time he saw her before the final exam as she was walking out the door. He tried to look her up in the university records, but on all her application and registration forms she had checked the little box that said "Do not divulge address and phone number." That was when he discovered that there are two columns of Olsons—over two hundred Olsons—in the San Francisco telephone book.

Well, he thought, she was a lovely exciting girl with a good attitude. I think she'll do just fine. "I think you're going to make it too, Connie Olson, and good luck to you." He remembered her term paper: "Yours Truly," Connie Olson. She's terrific, he thought, and shrugged his shoulders. This was what faithfulness to principle had cost him. Well, what the hell. How could they *be* principles if they didn't cost something?

In August his phone rang, and when he picked it up a young woman was saying, "Hi, Daniel? This is Connie. I thought I'd just call to check back in with you and let you know what I've been up to. I dropped out of school and I've gone back to modeling—"

"Modeling?"

"Yeah, and I'm doing really well—"

"Look, Connie, I can't really talk right now, but can we get together for lunch? How about tomorrow?"

"Sure."

The next morning, driving over to North Beach, he thought, With a girl this beautiful, Daniel, you've got to keep your cool. You can't be the eight hundred and seventy-seventh guy to tell her she's gorgeous. He met her in front of the restaurant. She came trotting up wearing a silky, clingy white blouse and a woolen skirt in beige. He might have pegged her as a young married from Hillsborough. But then he saw that even though her clothes and her carriage referred to that sort of understated elegance, her own rhythms, her young eagerness, did not. She kissed him on the mouth, there on Broadway, where he was deliciously aware of standing on Broadway being kissed by an amazingly beautiful girl. But it was the kind of kiss women celebrities will give Johnny Carson. It meant only Hello.

This was one of her favorite restaurants in San Francisco. It had been converted from a theater, its huge interior partitioned off now into separate dining areas, each one with a different motif—a leather-and-neon high-tech "Disco" with flashing lights, an Arabian tent with ottomans and carpets, a "modern" room with a huge skylight window and lots of ferns and bare cedar. But it had only one vaguely international menu, so that it was possible to have enchiladas in the Arabian tent and stuffed grape leaves in the Disco. They ate in an approximation of the sort of ice-cream parlor the Bowery Boys hung out in. As they walked through the lunch crowd, he was aware how the men around him would see her and then look again, but they yielded to something in her bearing and looked away.

"I used to model when I was about fourteen, but I didn't think I was mature enough then to handle it. But now, last spring, this old friend of mine who's in advertising that I used to know from when I worked at B. of A.—he has a girlfriend who owns a boutique in Sausalito, and asked me if I would do an ad for her. Well, the ad is in all the BART stations now. I didn't get paid or anything. I mean, she gave me some clothes and stuff. This blouse. But it was fun, so I checked around with various agencies and I'm in with one that I feel is a good one, and I'm starting to put together a really neat portfolio. Look at this. This is just four months' work—"

It was a black zipper-bound spring binder filled with mounted 8″ × 10″ glossies of her in standard modeling poses. In some he would not have recognized her. The poses ranged all the way from cowgirl in tight jeans to haute-couture ball gowns in severe, coarse-grained black and white. She could do the cold-as-Garbo beauty and she could do the young mother holding the baby on the Ivory Snow box.

"This is amazing," he said. "I wouldn't have recognized you in these. And these two don't even look like the same person."

"I know." She was smiling, nodding her head. "That's what makes me good. I can do any look the man wants. The only thing is I can't do stockings and some kinds of really high fashion, because even though I have a long torso, my legs aren't long enough."

"Aren't long enough? They reach all the way to the floor, don't they?"

"Right."

In one picture she was wearing a one-piece bathing suit cut down to her navel, her arms pulled back and up, her hands behind her head, her bosom lifted and firm. In another she was lying on her side, wearing only a lacy bra and bikini panties. "Lingerie pays more," she said, "because it's harder." In the next she was wearing a see-through blouse, her nipples plainly visible through the cloth.

"How would you go about putting together a portfolio

like this?" he asked her. "Do you have to pay a photographer or what?"

"Oh yeah, most of the time. But I've walked into studios and talked with photographers and gotten them to shoot me. I'm good. I'm a good model, and they know it's to both our advantages, so they say yes. I got the best fashion photographer in San Francisco to shoot me for nothing. Sometimes they say no, but I just leave them my card and tell them, 'I'm sorry you feel that way, but here's my card, because I think you may want to call me again in the next few days after you think about it. I think we could work very well together.' And they *do*."

She was looking at him directly head on. What photographer would resist the chance to call her? Her dark blue eyes were fringed with thick lashes. She believed in herself absolutely. Her mouth, her lower lip, was full, and made her look both sensuous and vulnerable.

"I've only been doing this four months and I've been in six ads and one fashion spread. I'll be on a magazine cover in Japan. I've signed to do some TV commercials. Photographers like to work with me. They know I'm good because I'm really uninhibited about my body. I know I've got a *gift?* and it's up to me to develop it. And I knew I had to do it now, while I'm still supple. If I had waited till I finished school, it would just have meant two more years that I couldn't be earning what I think I can. The sky's the limit in this business. I could make a hundred thousand a year. I mean, look at Lauren Hutton. Look at Cheryl Tiegs."

Daniel looked at her. Her nose was just on the verge of being too long. This monologue of hers might have been pathetic, as pitiful as a twelve-car model railroad in the window of a hardware store—a token of reality attempting to *will* itself into standing for the whole shebang. But somehow she affirmed herself. Her own belief in herself made her incandescent. She's perfect, he thought, incredible.

"My agent loves me because I'm absolutely mature and dependable. Some of these girls will get sent out on a job and they won't even show up. But I'm always there early enough to do whatever makeup the man wants me to do before we shoot—if he wants me to be 'real wholesome' or more like 'high fashion' or 'young executive.' I do all my own makeup, so I don't need a makeup man. I didn't think I was mature enough before to handle the pressure and the rejection. But I sure am now, absolutely professional."

"Why do you want to go putting such a stress on maturity?" he said, smiling. "You shouldn't be in such a hurry to be mature. Why don't you stay *young* for a while? Enjoy yourself—"

"Oh, I enjoy myself. But I'm going to make it too, and you've got to be mature for that. You can't let things that get you down interfere with your professional life. That's what I mean. I want to be absolutely professional. Like, last week I had this huge fight with my boyfriend just forty-five minutes before I had to leave for a shooting in Tiburon. I took a shower, did my hair, got my makeup stuff together and got there—*on time.*"

"Why did you mention your boyfriend just then?" he asked, still smiling.

"What do you mean? That's the way it happened."

"You were giving me an example—why *that* example, the one with your boyfriend?"

"I don't know, it was just the one that came to mind."

"You wouldn't have mentioned him just to sort of let me know you *have*—or *had*—a boyfriend?"

"It was just the first thing that came to mind, but—"

"You wouldn't have let me know about the boyfriend because you were concerned I might make a pass at you?"

He was still smiling, leaning back in his chair. She was smiling too, but she was less at ease now. "Were you concerned I might make a pass at you?"

"Well—"

"Would you be disappointed if I didn't?" She was beginning to color just a little now. Her skin, the skin of her face and down her neck, was *incredible*. "Would you like to have an affair with a married professor?"

She smiled and said, "I just called to get back in touch with you and let you know how I was doing. I— *You* asked me to have lunch—"

"Well," he said, taking out a pen and paper. "Here's my card, because I think you may want to call me again in the next few days after you think about it. I think we could work very well together."

In Bremen, Indiana, they stayed at the Hoosier Lodge, where Danny saw his first Amish people driving a buggy down the highway. In the morning, shaving, David thought of Daniel and how he had shaved every morning in the house with Annie during the past two years while he was having the affair with Connie. What did he think as he shaved the face he saw in the mirror? This man is loved by two women? This man keeps two women happy—forty-five and going gray as he might look? This man is enjoying his life? This man is a liar?

You had to get out of the house to have an affair and you had to cover your tracks, which meant you had to tell lies. The lies had to be convincing, yet have all the air of thoughtless talk, and they had to be consistent, the whole story hanging together. The lies had to be packed with details—times of day, names of people and places, specifics of money spent and change left—yet they could not put on too much of an air of prefabrication. They could not sound like alibis, which of course they were. In these alibis you could get away with *some* use of actual people known to Annie, but unless you were ready to enlist those people openly—and Daniel had kept the whole business secret from everyone for two years—you had to keep that sort of thing down to a minimum. And that meant you had to begin inventing people with whom that time was spent.

Getting away during the daytime was no problem: the structure of anyone's day has any number of vacant spaces that aren't even apparent from the outside, and so aren't even questioned—between classes or afterward. Getting home half an hour late, or even an hour, could always be blamed on traffic. But how to get away in the evening, or even for a whole day? Daniel thought of teaching some evening courses, but then he would have to explain what he had done with the money he earned from them. He began to check out the *Evening Division Bulletin.* The courses had to have some plausibility for him, yet be absolutely uninteresting to Annie. "I was talking to Grierson today—you know, the librarian—and he was telling me how more and more research stuff is being computerized for better access. Like, all the articles on Hemingway that discuss food. The Evening Division is offering a course on beginning programming on Wednesday nights." "McDowell is offering a Philosophy of Law course on Thursday nights—Theory and Structure of Criminal Law. Sounds like fun, do you want to sign up with me?"

Then these courses would have to be peopled with students and with an instructor who was identifiable by a repeated mannerism, like a character in Dickens. Within the classes rivalries would break out, personality clashes would develop, young kids would put the make on each other or fall in love. "This couple is really cute. They really think their whole thing is a big secret, that nobody knows." He found he even had to do some research, to learn enough computer jargon to convince a person totally outside the field. "McDowell" was very dry, very *remote.* "Very straight. Right wing—N.R.A., anti-abortion, the whole shmeer. Cadaverous, with a big beaked hook nose on him like that, like Dick Tracy. But also very sharp, you know, like Buckley, complete with the eyebrows that go up and down like that—'I mean, *really.*' "

One day, he did meet Grierson in the library. The man was carrying an athletic-equipment bag with a squash racket sticking out of it. "Where do you play?" David asked him. That night he told Annie he wanted to play squash. "Running is great, and I like it that we do it together, but it doesn't feel competitive enough. It doesn't have enough of an *edge.*"

"Sure," she said. "*Quien es más macho*, right?"

"Oh yeah, what the hell."

Three afternoons a week he ran with Annie up and down the hills of Dolores Park, and two evenings a week he "played squash" at the Telegraph Hill Athletic Club. It was perfect because now he only had to invent two or three squash players and snippets of their conversation and maybe some scores and a few details—"He always plays to my backhand, tight up against the left wall, and I always end up hitting my racket against the wall. But I'll get him." The sport justified the shower he took to wash Connie's perfumes and cosmetics off, and it let him claim to be exhausted if he felt like it. If he'd been drinking or smoking with Connie, he could always say he had gone and had a drink afterward or had a couple of tokes in the car on the way home.

"You sound like you're getting pretty good," she said one night when he'd been going on about his game. "I'd kind of like to watch you play."

"Annie, it's a *men's* club. But if you want to learn I can switch over to a co-ed club."

"No, it sounds too indoorsy. But don't let those gay guys get too chummy in the shower."

"They pretty much keep to themselves."

The challenge became not only to prolong his times with Connie, but to make those times routine to Annie, so that they were accepted without question. Taken for granted. And he found now that he needed that extra time, and felt that Connie needed it too. Not just to fuck, but time before making love, and afterward, or just to be together without fucking at all, an evening in front of the fireplace, or watching TV, or going over her portfolio. And it was during one of these evenings when they were sitting on the rug with their backs against the couch, watching the rooms the fire made and the shapes moving around within those glowing rooms, that he let his head fall back in the fullness of his feeling, let his head fall back and feel the nubby fabric of this beautiful girl's couch on the back of his neck (the pressure of her thigh against his), and looked up at the stippled white ceiling with all those bits of sparkly, mica-like specks all over it. For a long moment he forgot the girl, so absorbed was he in feeling the distance between the top of his head and the wall on the other end of the room, behind the couch, feeling the volume these walls enclosed, the pictures Connie had hung on them, insisting that the apartment become part of her life, the wall she hung her pictures on. And yet the walls had been here long before she ever saw the place—the rooms laid out and named—living room, dining room, bathroom—insisting on their function, oblivious to her. They would be these rooms to anyone. But they were *Connie's*, he thought, feeling actually petulant and realizing as he was doing it that he was kicking his foot out once, sharply, as if stamping.

"Hunh?" Connie had said.

"Nothing," he said, squeezing her shoulder. But the thought bothered him enough that he found himself letting himself into her apartment one afternoon when he knew she was in L.A. on an assignment. He walked slowly through the apartment, standing for a long time in these rooms and areas, every one of them, slowly turning his head and even turning his body this way and that. Connie's living here had not gone deeper into the apartment than her choice of the paint and the wallpaper. But these alone—and the furniture she had chosen, and the pictures, the candlesticks from Mexico, the spoon that said NEW JERSEY on the table in front of him, the small plastic syrup bottle in the shape of a bear still standing on the kitchen counter where she'd left it this morning—all the barely perceived details of her life mutely proclaimed her presence, which now he felt as a buzzing charge filling the air of these rooms. He lay down on the rug in the living room, tasting it with every pore of his body.

That night he told Annie how excited he was about the prospect of going fishing with some friends of Lyman's. Four days in the foothills of the Sierras, up in Calaveras County but beyond the Mark Twain country. It was a remote canyon. You had to have a four-wheel-drive vehicle to get in there. You took the jeep to the top of the ridge, at about seven thousand feet, and then you hiked down thirty-five hundred feet in a matter of some four miles. It was a tremendously steep grade, and you were carrying forty- or fifty-pound packs and walking in dry streambeds where the loose rocks were always threatening to break your ankle. Then you had to ford the Mokelumne, still carrying your pack, and sometimes it was deep enough and fast enough to knock a strong man on his ass. "I don't think it would kill anybody but you'd swallow a lot of water and get banged around on some rocks pretty well before you got back on top of things." Perfect.

He invented the landscape. "The slope going up to the

ridge is easy. They're like mountain meadows covered with this *miner's lettuce?* It's a big plant that looks like a cross between a cabbage and a philodendron. Then when you start down into the canyon itself you realize you're in ski country without the snow. The trail goes right past the top tower of the chair lift. The trail down is grueling but it isn't dangerous—except I guess you could break an ankle. And it would be a bitch carrying you out of there if you did.

"The campsite itself is a dream: a white sand beach at the foot of a rock outcropping that goes up twenty feet or so. And right on our front doorstep is a good-sized pool. Albertson has gotten a couple of twelve-inchers out of there, but I've never had any luck. A thousand feet or so upstream the river comes out of a hole in the mountain like a fire hose, and this enormous granite apron is spread out at the foot of that cliff, and the water has been damn near polishing it for about five million years. Right there it's all spread out, it feels like it's going ninety miles per hour but it's so clear—it's only two or three inches deep, just like a moving *sheet* of water that never stops flowing over that granite."

He invented the men he went with. He took men he knew and gave them different names and took them up to the mountains with him. "Duncan's trip is rattlesnakes. His idea of machismo is pinning a rattlesnake with a forked stick and then snapping pictures of him with his free hand." "Walking through the woods with Black—he says, 'I'll catch up with you.' He pulls a wad of toilet paper out of his pocket and waves it at me, saying, 'I've got to stop here and *vote.*' I love it." "And when we get to Davidson's studio to pick him up—this is like seven a.m.—he's standing in front of a canvas tacked on the wall, painting away like a madman, except that he's got no clothes on—not a stitch. Anselm takes one look at him and says, 'Pretty kinky!' And Davidson looks at him and says, 'I can explain everything!' "

One day when he had just "gotten back" from one of these fishing trips, Annie asked him, in all seriousness, "But don't *you* ever catch any?"

"Sure," he said. "I've caught—you know, a fair amount. I mean, I don't have the kind of experience these guys do. But we have them for dinner right there. Do you think I'm Ernest Hemingway?"

"Why don't you bring some back, and we can have them for dinner here?"

When he "got back" the next time, he said, "Here you go. I brought back my whole catch." And then he plopped the two rainbows on the kitchen table. They were still lying in their Styrofoam tray and wrapped in plastic from the Safeway. They both had a good laugh over it. Annie said, "Figuring gas, equipment, food and liquor, those two trout couldn't possibly have cost more than fifteen or twenty dollars a pound."

In Iowa City they stayed at the Ironmen Inn, a place that looked like the *Führerbunker*, the walls plastered with old news photos from the heyday of the local football team. Was it the University of Iowa or Iowa State? One had a football factory and the other had a writing factory, and he could never keep straight which was which. Danny loved it—the place had an enormous indoor pool with a Jacuzzi attached—but it was a mistake. The motel was *miles* from town, and they had to eat in the Ironmen Room. As they grumbled over dinner, Danny said, "Well, look at it this way: the pool is open twenty-four hours."

As David was showering before dinner, he thought how pleasant it was going to be being back in their own house.

Just the smallest luxury, like knowing where everything is, or actually *liking* the pictures on the walls, pictures they had chosen instead of the "Utrillos" outside that might have been painted and sold by the yard. Being in the house that was their own, where they had their lives, was such a deep pleasure, one he had never expected himself to claim, or to savor this way, in anticipation, remembering how he could sit in his studio "composing," but leave the door open so he could hear Jane noodle around on the piano on some old Billie Holiday tunes. Then maybe Danny would come downstairs and just happen to have his harp in his pocket, so he would chip in. As he soaped himself, he wondered how he could make more time for that sort of thing when the fall term began and both he and Jane had to start their teaching schedules again.

And then there was Daniel. David had always assumed that Daniel was simply a part of that music that made up his life. And yet he must have simply *not been there* so much of the time when they'd been together these last two years. Out to lunch. It challenged the hell out of anybody's belief that internal states manifested themselves in external signs. Had he really loved *both* women, so that he got some kind of satisfaction with either? There must have been times when he was with Annie—hell, when all four of them had been together—that he would have given almost anything to be quit of this whole scene, to be with Connie, where he was not required to continue this exhausting pretense, living this fantasy of someone else's devising—marriage, dinner parties, concerts, plays, weekends in the country. Amusements for the middle-aged middle class. They went to a prize fight one night at Kezar Pavilion, which was pretty pathetic, pretty boring as a fight—except for the beer-swilling redneck behind them who kept yelling, "Make him bleed!" *He'd* been enjoying it, why couldn't they? Amusements for the middle-aged who wanted to go slumming, David thought as he worked the shampoo into his scalp.

For Daniel, being in his own house would have become not a torment but a form of non-time, a time not fully engaged or lived, like the page of a book that when you got to the end of it, you asked, "What the hell have I just read?" *Downtime.* Daniel would be conscious now of the spaces of the rooms in his house, of the air they contained, the spaces they measured out to such precision—and for what? This could go on indefinitely, and in the beginning he thought it would, the house and Annie and his marriage to her providing a framework for the time he called his life, within which he could *get away*, escape that framework or introject a totally different structure into the interstices of this one, a secret room in which he began now more and more to have his true life, the life in which he was truly excited, truly generous, truly passionate, truly alive. Being in the house was getting more and more to be like what his sister-in-law said about someone stupid: "Lights on, no one home."

And then there was the delicious guilt of the alibis and of the transgressions themselves, the challenge of the detail. He spent half an hour banging the frame of his squash racket against the floor and balustrade of Connie's terrace. He had scuffed up his gym shoes the same way. He would put on his squash outfit and then run in the Presidio with Connie, getting the clothes good and sweaty before bringing them home and throwing them, casually, into the wash. This, he thought, made a lot more sense than the evening courses he had "taken." Then he had actually had to read some of the crap on the reading lists, books on Basic and Pascal, or R. Charles Mortimer's *Gens and Justice: The Law in Early Roman Civilization.* What a horrible, constipated style the man wrote in!

But part of the pleasure was just in this breaking of the

law, seeming to operate within it while actually saying "Fuck you" to it, so that everything he did was charged with this doubleness, knowing something nobody else did, having the being *they* thought was him at the same time that he had this other self, this secret being asserting itself, unseen, right under their noses. The sweetness came in part from saying "No" to the very system that seemed to provide the framework for his life, what everyone simply assumed he had accepted too, just like them.

Daniel regretted only that all this acting, this performance, was going unacknowledged. He was his only audience, and that annoyed him. A job of acting presumed an audience aware that the actor is playing a role: it was validated in that audience's knowledge of its doubleness, its approximation of an identity. When the actor's self fused with the character's, there could not be any consciousness of similitude and no appreciation of the details chosen to create that illusion, whose success, he now realized, depended on its being unsuccessful, on its being acknowledged *as* an illusion. Imperfect. Walking back the two and a half blocks from Connie's to where he'd parked his car one night after one of his "squash games," with the feeling of her still clinging to his skin like the briskness of a shower on a cold morning, he realized the studied casualness with which he was walking, as if he lived on this street, in this neighborhood, taking it for granted, not looking around while he walked along, "lost in thought," conversing with himself, even. The worst was one morning at breakfast when Annie showed him a newspaper ad for an extraordinarily sexy "teddy." "Would you like me in something like that?" she asked. He looked at the ad and swallowed hard. It was Connie.

"Sure," Daniel said.

As he soaped his genitals, David thought that because the self behind Daniel's role-playing went unperceived there must have been times when Daniel began to feel the reality of his *being* leaking into that role. He was saying "No" to the system, the structure of middle-class middle-aged couples, but the negation was so all-encompassing it negated eventually even the self that made it. At first, when this doubt about the reality or fullness of his own being began to trouble him, he could find comfort in his affair with Connie. Sitting at breakfast with Annie, the litter of dishes and crumbs and sections of the *Chronicle* insisting on the messiness of the real, buttering a slice of toast and making sure the butter went all the way to the edges, he would feel himself begin to dissolve, to be aware of the room as if he himself were not there, the walls and the windows, the floor and the ceiling shaping and containing this space and these smells so completely that *he* faded into a kind of irrelevance, like the crumbs in the bottom of the toaster. If he smoked, he thought, he would right now crush the butt of his cigarette brutally into what was left of the cube of butter as it sat in its glass dish there in front of him.

He would think then of his previous night with Connie and the perfection of their being together, her breast just *there* in his hand, its nipple just manifesting itself through the lace, his cock being sucked *just* to the point of spasmodic surges of pleasure and then allowed to ebb back, as he had brought her, repeatedly, to the *edge* of coming, both of them asserting their desire and their control. That had been real, he thought, that had been full.

And then he realized that he had made those very motions, he had spoken the same words, to his wife. And Connie did not know that she too, the person he had chosen or been given to share this most secret room of his life, stood outside it, unaware of any illusion and so unable to appreciate the reality of his performance *as* a performance. Toward the end of the second year, lying on

Connie's bed or sitting with her at breakfast—one morning she did cook breakfast and served it and ate it with him while wearing nothing but a little lace apron that frankly failed to conceal her breasts and her pubic hair—he would think of this apartment as a place measured out, as the rooms of his house were, to accommodate a life, and that this life only *happened* to include him, and only during the time he was actually here. Twice—no, three times—he took things from her apartment, nothing that had any value—a ceramic matchstick holder from Paris with CASSIS QUENOT printed in big red block letters around the base, a little toy metal seal, a glass ashtray with something printed on it in Arabic. But outside her apartment they seemed to lose whatever charge they'd had. He left things of his own there, too—clothes and keys, a small spike from a narrow-gauge railroad in Colorado that he'd had brass-plated. But none of these things made his being in this place any more necessary, and he felt a strange ghostliness as he moved through the plushes and glass and chrome of this apartment.

The ghostliness was similar to what he felt as he drove away from his own house to "play squash" or "go fishing." The elaborateness of the cover story made him feel like a criminal. This is what criminals must feel like as they prepare to do a job, he thought, constructing a world based on the fullness (and falseness) of the cover story. And yet he was not going to commit a murder or rob a bank or even burglarize a house. He was only going to do something so *normal* the wonder was that it did require such an elaborate preparation. But it was just that combination of secretness and commonness that made it so sweet. It was what everyone wanted and almost nobody did, to slip out of or through the structure that gave your life a shape into a room where your life took the shape you wanted it to have, to love and be loved by someone perfectly beautiful. The dream of millions. He could never have predicted either that he would achieve it or that, having it, he would find it so fucking complicated.

"Hey," Jane said, poking her head in the door of the bathroom, "you must be the cleanest man in Iowa by now! Should we just go on down to dinner without you?"

In the Lone Pine Inn in North Platte, Nebraska, David thought how Daniel must have wondered how these two women had their being while he was not there. He imagined Annie making dinner by herself, or stacking the dirty dishes in the machine, or repotting plants out in the garden, humming to herself as she worked, feeling the cool redness of the flowerpot as she held it in her left hand, tamping the soil down around the stem with the trowel in her right. She would talk on the phone with Andrea. She would talk on the phone with Hendricks from the California Foundation. She would walk through the quiet house turning off lights on her way to the bedroom, taking the whole house for granted, like a second skin and not the occasion for doubts or anxieties about her own being. She would watch the late news, and then she would read for a while, her mind quiet, her life comfortably full. Even Daniel's absence was part of that comfortable fullness. And Connie? Connie would work out at the gym, and then she would drive home because she preferred to shower in her own apartment, where she could reach out through the plastic curtain and have a hit or two off the joint in the ashtray on the windowsill. She would turn on the TV but she would not watch it, except when the commercials came on. Instead she would sort through the latest batch of glossies to see if she wanted to update her portfolio. There were some nudes in this batch, and even though she knew she could command a higher fee doing fashions, she toyed with the idea of including one or two, just to put a sort of *edge* on the portfolio, to make it a little more exciting to the men who would be checking it out. The phone would ring and she would talk to her mother for a while, thinking that she would ask Daniel about the nude shots. She would not be aware of the color of the walls or the texture of their stucco, or of the starburst chandelier over the table in the dining room. Whenever they had dinner at home, Daniel always wanted to eat in the dining room. Why was that? she wondered. It was so much less of a hassle to eat in the kitchen.

In the Maybell Lodge in Maybell, Colorado, David took the little square plastic bucket from the room and walked down the sidewalk past the other doors and windows to the ice machine. The metal scoop had been left in the ice bin and its handle was uncomfortably cold. This was the first motel of the trip home that actually had a separate little bedroom for Danny instead of just a single bed on the far side of the room, and he was savoring already the lovemaking he and Jane would do—for the first time since starting home—later tonight. He put the ice into the glasses and poured the whiskey. Danny had turned on the TV and was busy watching the latest installment of the rerun of *Roots*.

"This is the first time," Jane said as she took the drink, "that I feel we've really gotten back to our own territory. As soon as I saw the Rockies, it felt like we had *finally* gotten out of the—the gravitational field of the East and the Midwest. Know what I mean?"

Behind the boy's back, David reached over and caressed her breast, while she gave him a big smile and stroked his thigh once, a little uncomfortably, and then pulled his hand away, murmuring, "*Plus tard.*"

He thought of Daniel and Connie as they browsed through her portfolio, or through a new batch of glossies or tear sheets, trying to decide what to include and what to cut. They came across two lingerie shots in a row. In the first she was wearing a sleek, floor-length gown in some dark, satiny material. It was completely opaque, but cut so that it showed perfectly the swelling roundness of her bosom and the shape of her nipples. The background was dark, with three tall white candles looming up mysteriously behind her. She was holding a wineglass and looking at somebody who was just outside the frame of the picture. The skin of her shoulders and arms and her swelling bosom was luminous, and there was no mistaking the knowledgeable sexiness in that look. The man this nightgowned woman was looking at was about to get the fucking of his life. This woman would show him that he only thought he knew what it was all about. Her look and her carriage, even the way the wineglass gave itself to be cupped in the warm flesh of her hand—all spoke of velvet-lined chambers in châteaux outside Paris, of satin sheets over the chaise on the terrace of the palazzo at Fiesole, looking down on the Arno as it wound its way, lovingly, through the body of Florence.

In the other picture, she was standing in a field of flowers wearing some sort of gauzy, flowered shift or baby-doll nightgown whose almost perfect transparency wrapped her—incompletely—in a cloud or halo of light. It draped down from her bosom and opened in front, exposing part of her upper thighs, but what was covered could still be seen through the sheerness of the cloth. The soft focus of the picture blurred her nipples somewhat, which seemed now so clearly visible but the next moment looked as if they were actually part of the floral design of this chiffon-like stuff. The pale sunlight that picked out the flowers amid which she stood was also caught in her tousled hair.

This was a world of morning sunlight and morning flowers, of the first awakening to love or the first awakening from the discovery of love and its sweetness—the look in her eyes that ambiguous.

"Do you like that?" she asked, her eyes bright with desire.

"Un-hunh!"

When she came back, she stopped in the doorway, leaning a forearm against the jamb, which framed her. She had tousled her hair and put on a gauzy, shift-like gown in the same style as the ad, a flower-printed "dress" he could see through that was loose and clingy, and widely slit up the front, that made a sort of zone or field of semi-transparency within which her long, lithe body moved subtly. She was giving him the same look of hungry innocence as in the photograph. She was incredible!

"Oh, honey," he said.

It started out as simply as that. At first she asked him, "Would you like me to wear something special?" Then later it was "What would you like me to wear?" Still later she would alternate between asking him and surprising him, either putting on an old favorite or wearing something completely new, an outfit he had never seen before. At first it was a particular nightgown: the long pink one with the demure bodice and the tiny little sleeves that he liked to slide off her shoulders; the floaty black chiffon in several layers; the bride-like gown-and-peignoir with an Empire waist that wrapped her in an obscuring white cloud of innocence, and that, when he'd peeled it off her body and her arms and pulled up the sheer white of the gown above her thighs, revealed a *wicked* white G-string bikini that cupped her blonde little pubic mound; the dark blue outfit that started with a slip slit up to here, which she eventually pulled over her head to stand before him in a matching bra and panties that led up to those moments of truth when she stood or knelt, her hands coming together at her sternum as if to pray, and unfastened the bra and pulled it open to expose her perfect breasts and then hooked her thumbs with deliberate casualness in the elastic of her panties and pulled them down slowly—revealing more and more of her flat abdomen, and then her most secret hair. Sometimes she would wear nothing but a garter belt and stockings, but she was carrying a wine-colored peignoir. By maneuvering the robe and her arms and legs, and by turning her body this way and that, she could be with him like this for a long time without ever actually revealing herself to him except in momentary glimpses. His part was to sit back and "enjoy the show," but after a half hour or so of this he was pulling the cloth out of her hands and burying his face in her bosom. Once she pushed this routine just a little further, standing beside the couch with her nude backside to him and spreading the robe over the sofa. When she did this, he could see her naked front in the mirror on the far wall, and he took her from behind, her rear end draped over the armrest while he stood between her spread haunches, his pants spilled around his shoes on the red kilim, fucking her in the ass while she hugged the cushion to herself and whimpered like a dog with pleasure. Once he stood in the doorway of the bedroom and watched her bring herself off with a penis-shaped hash pipe.

Then they progressed to street clothes. He loved to watch her undress, so she became a Montgomery Street secretary getting home from work and slowly pulling off her raincoat and scarf and sweater, and then, prolonging each step while he stood watching her from the doorway to the bedroom, she would slowly, with excruciating casualness, unfasten the silk bow of her blouse and begin to unbutton it, pausing to gaze out the window as if preoccupied with something, or to talk on the phone, all this time her unbuttoned blouse open to the waist, revealing her perfect, lace-covered breasts. She would go on with this, pausing to look at herself in the full-length mirror. Or she would wear a cocktail dress with nothing at all under it, and in that elegance, in jewels and makeup, she would slowly sink to her knees in front of him as he leaned back against the back of the couch, and suck him off. In another version he had her buy a nurse's starchy white, semi-transparent uniform, under which she wore enough underwear to keep her busy undressing for half an hour—a camisole and half slip, a full slip, a body stocking, two bras and three pairs of panties. "Daniel," she said as she came out into the living room for him, "I can hardly move."

She would do a cowboy shirt and blue jeans or a tight satin sheath with spike heels. They fucked in the kitchen and the bathtub, starting out in everything from jogging or tennis outfits to an incredible floor-length knit something she had gotten at Helga Howie for an evening at the opera. Sometimes she would start out naked and stalk him throughout the apartment, stripping the clothes from his passive limbs after she caught him, and then she would stroke him and suck him till he begged her to let him fuck her.

Later on he got involved in the costuming too, and in one of their favorites she would greet him at the door wearing a simple cotton shift, like a farmer's daughter, and he would do a "coke dealer"—a black silk shirt, an Australian outback hat and Frye boots. He would spend all evening seducing this innocent young thing, eventually talking her into snorting some cocaine. At that point she became a drug-crazed, sex-crazed hellion and began ripping his clothes off. In addition to the nurse's uniform she had a short-skirted, candy-striped thing with a pinafore-type bib like a waitress, a nun's habit and an olive-drab WAC outfit complete with the little cap and some campaign ribbons pinned over her left breast. The best one was when he played Uncle Daniel, who had come to visit her mother, who was out right now. Connie was a fourteen-year-old virgin who had just come home from the convent. For that one she started out wearing a version of a Catholic school uniform: white blouse with a navy cardigan, a plaid skirt that barely came to her knees, white knee socks and black patent-leather shoes.

" 'Hi, you must be Connie. I'm your Uncle Daniel. Is your mother home?' "

" 'No, Uncle Daniel. She went out with Uncle Malcolm. She told me she wouldn't be home all night and I have to stay here all by myself. Would you like to come in for a glass of lemonade?' "

" 'Why, I'd love to, Connie.' "

Perfect.

"Do you think anyone actually *lives* in Ely?" David asked as they walked along its main street in the gathering dusk, trying to choose a casino in which to lose the ten dollars apiece they had allowed themselves.

"Sure," Jane said, "and some of these people call it *home.*"

The air of the West felt different—thinner and drier, he thought. There was a difference too in the openness of the perceptual field. When you looked, as he did now, straight ahead into the failing light of late summer evening, you saw ground and buildings, a distant horizon with maybe some craggy mountains to mark it and then, above that, the sky. In New York City your perceptual field was absolutely jammed with city—with buildings and people, cars, the whole cacophony of urban geometry. There was no horizon. In the Hudson Valley and the area around the Taconic Parkway, even as far east as the Berkshires, the horizon was always pushed near—a succession of broad, shallow valleys. But here in the West the sense of a nearly horizonless openness and possibility was overwhelming. Some people could call these literally wide-open spaces home. But not he. He needed a sense of structure and of limit to play against.

That was one reason it felt so good to be actually coming home. For most of this summer they had been away, staying at friends' or relatives' houses on their trip East, and then for two months at the Mortimers', and the prospect now of being in his own house again was almost palpably sweet. To stand again on the stairs, looking down into the living room as the afternoon light came through the tall casement windows, and feel the spaces of the house as they enfolded him and his life with all the details of his and Jane's choosing—the bird kites on the wall, the neon cocktail glass the Riordans had given them, which they turned on now mostly only for parties.

The Riordans. Daniel. Daniel, it seemed to him, had always enjoyed the house that he and Annie had lived in now for—what was it—seven years? It was not a big house, but the spaciousness of its rooms made it comfortable, the fact that you could look all the way through the house. From the deck that Daniel had built overlooking the backyard with its functioning vegetable garden, you could look straight through the dining room and living room and out the front bay window that looked down on Dolores Park, and beyond that, the city, and beyond that, the Bay. Daniel had built the deck; he had terraced the backyard and brought in sacks of topsoil. He had terraced the front yard and consulted with a landscape architect about how to plant it. He was always involved with projects—painting or stripping paint or building shelves or organizing storage space in the basement.

During the past two years, had Daniel stood in the kitchen picking shreds of meat off the carcass of a chicken at two or so in the morning when he could not sleep, and had he then felt the self he had, the self that was his as it was conscious of the spaces of these rooms, the birches outside the window over the sink, the funky old ceramic dish David and Jane had given them that said "The road to a friend's house is never long"? Had he sat in the dark living room while Annie was asleep upstairs and quietly clinked the ice in his glass to drink a toast to this city whose lights now gave themselves to his view? Had he gotten up with the morning light flooding the bathroom and been only half aware of its tiles?

Or had he had that being with Connie, to drift off to sleep in her apartment, wrapped in the knowledge of its sliding glass doors and their aluminum tracks? Had he looked at his own face in Connie's mirror when he shaved in the morning, taking for granted the tub behind him with its clear plastic shower curtain bordered in red to look like the cover of *Time* magazine?

And her, Connie (was that Constance? or Cornelia? or?), what knowledge did he have of her? In *The External World* he'd written a story about a strange man who wandered through the stacks of the library, taking down books at random and reading only their dedication pages: he'd never admitted to anyone that the man was himself. Connie now was like one of those books that that man looked into as if he were looking into an office that was someone else's, seeing through the partly open door only a thin slice of the end wall and part of a couch covered with an afghan. What did he see of her when she had taken all her clothes off? What did he taste of her in their most abandoned lovemaking, when she tilted back her head and closed her eyes, leaving him lying there looking at the skin of her face and her eyelids?

With "Uncle Daniel" and the "farmer's daughter," they had been inspired. Her apartment had been their world. They would run together in the Presidio, but there was almost no place else where they could be absolutely certain of not meeting someone who knew them—or him. They did not even go to movies together at first for fear of that. It was only close to the end of the first year that they began checking out the movie listings in places like Daly City, Orinda and Fremont. Then, on one of his "fishing trips," they took the silver Amtrak train down to Santa Barbara for one of her modeling assignments. The art director wanted to use some tony resort down there as a backdrop, and then she would tape a commercial for tampons. They stayed at a wonderful beachfront motel and flew back. The ride down was something quite new. There were other people around. This was the outside world, and they were out in it.

Daniel could not remember anymore whose idea it was, but somehow it got decided that they would separate and then he would "pick her up" in the club car. After she left him, he looked out the window. What town was this they were passing through at half speed, some bell clanging patiently somewhere, that seemed emptied of its people in the gathering dusk? They passed a post office, but he could not read the sign. *Coming Home* was playing at the local movie theater. Laundry was hanging from the clotheslines in the backyards. As they passed a school yard, he saw two kids playing basketball: one of them jumped and sank a perfect twelve-footer, the ball touching nothing but the net. At the edge of the woods, in the front yard of a large white house with many roofs slanting at odd angles, a young boy was shooting a bow and arrow at a target Daniel could barely see in the failing light. The sense of space was so engrossing that for a moment he found he had even forgotten Connie, and he jumped to his feet.

He sat down at her table in the club car and put his hand on hers, wanting to tell her how he felt, how the long, moving, constantly swaying space of the train had affected him: a kind of nostalgia or dislocation in which suddenly he felt closer to her than ever before. *This* was love, he knew now, more than his love for her beauty, for her perfection, but a love for the *place*—hard to describe in any other word—the zone or area that their being together like this in the outside world, actually moving through that world, created. It was the first time he had thought of using the phrase "our life." "This is our life, Connie, and I love it." He felt his eyes well up with the intensity of his affection.

"Pardon me, mister," she said, pulling her hand away *huffily*. "I don't even know you!"

They decided to check out the singles bars around town because no one he knew would have dared to go near one. At first she would go in alone and let herself be approached by single men, whom she would put off, one by one, while he watched from a seat behind some ferns at the other end of the room. Then, while the rejected dudes were still milling around, he would go over and make his move. She would melt for him and they would leave together while the scorned men eyed them surreptitiously.

He began to see her picture around town, and once, on the wall of the Montgomery Street BART station, she was looking directly at him from across the tracks holding an enormous kosher salami. Something in Hebrew blazed out at him in red characters from the wall beside her perfect face. He had no idea what the writing said, but he looked into her eyes and looked at the way her hands held that salami and he felt the full weight of his happiness.

Their final wrinkle on the singles bar was for him to go in first and get a seat at the bar. Then she would come in, sit alone and let herself be approached. On a busy Friday night when she wore a satiny cocktail dress with absolutely nothing on underneath it, the weight of her breasts moving visibly under the cloth and her nipples just there, she had four or five men hitting on her at once—an account executive, a young engineer in a woolen shirt, a redhead who owned a string of hardware stores (probably bullshit), an Arab who said he was studying architecture at Berkeley. She let them fall all over her while Daniel "ignored" them. Suddenly she spotted him. "Who the hell is *that?*"

"Who?"

"That man at the end of the bar, with the salt-and-pepper gray hair?"

"Him? He's *old!* That man looks to me like he's at death's door. And too weak with age to knock. Think he can get it up? No way, José."

"What do you want to bet he'll get it up when I lick his balls?"

Silence.

Then she would go over and fawn on him till he disdainfully allowed her to pick him up. Daniel would put on his raincoat *languorously*, looking over the field he had just vanquished.

They stopped doing that after the night they enjoyed it so much in Henry Africa's that they took the same act to Perry's on Union Street. One of the young guys Connie had rejected at the first bar showed up at Perry's, asking, "Hey, what the hell is this, *Candid Camera*?"

What the hell *was* this? David thought now. Where the hell was it that his friend had his being? Which of the faces Daniel presented to the world was his real face and how the hell did he find that reality acknowledged or affirmed? Where was it that Daniel *actually* lived his life? And there, driving across the flat open desert strewn with tumbleweed, David wondered also, as every moment brought him closer and closer to his home, where he had lived for ten years with Jane and their son, where he created his music and paid his bills, and ate his meals and dreamed his dreams, where he felt hungry and made love with his wife and jammed blues with her and Danny, where he had stood at any time of day and night tasting the quality of the spaces of those rooms, by himself and sometimes even when people were over, either for dinner or to play, those high beamed ceilings, those quirky alcoves—where was it that he had his own being?

The hard part about getting home to their own house was not learning to work the new burglar-alarm system, but adjusting to the fortress mentality it implied. The bell box (plainly visible from the street) and the signs and stickers all announced affluence and distrust, a paranoia that everyone accepted as simply being realistic. "Well, you've got to do something." And the fortress reasserted itself every time they left or came home, when they had to arm the system or turn it off with a little key that now of course each one of them had to have. If, when they got ready to leave, they had remembered to close all the relevant doors and windows, an LED on the panel showed them a green light. If they weren't getting that green light, they then had to go over the whole circuit—upstairs and down—till they found the door or window that was open and then closed it. Then, holding the front door open, they had to turn the little key in the control panel, which now showed them a red LED, indicating the system was armed. The alarm then gave them twelve seconds to get out and close the door. Coming in was the same story in reverse. As soon as they unlocked and pushed open the front door, they heard a high-pitched whistle that gave them twelve seconds to find the alarm key, insert it into the panel and turn it. Once or twice they just didn't move quickly enough or dropped the key or had to shuffle bags of groceries around, and before they could get the key into the panel their twelve seconds were up and they heard what a burglar would hear if he broke in—a breath-catching clangor of bells and sirens that they were told could be heard a block away.

"Good," David said. "I don't want a silent alarm that will allow the cops to sneak up and catch the guy. I want a really loud bell that will scare him *away*. I just don't want to come home again and find some dude coming out of my backyard with my tape deck in his hands. God knows we haven't got much—outside of the equipment in the studio, I think the alarm system itself cost more than anything we've got. It isn't even the property, really, since the insurance will replace that. It's the fact that the guy *could* pull a knife or a gun—and even beyond that, just the horrible feeling of being *violated*, of having your territory, your life, invaded like that."

In the pile of mail the Jerk family had left for them was a letter from the Career Criminal Division of the D.A.'s office telling them that Albert Boone had pleaded guilty to three counts of burglary and one count each of car theft and carrying a concealed weapon, and had been sentenced to a series of five two-year sentences, which were to be served concurrently. In two years—maybe less—the son of a bitch would be out on the street again. They also had to complete their dealings with the insurance company about replacing the receiver and headphones and some turquoise-and-silver things Jane had bought years before in Albuquerque. Then David had to take the turntable in to get the cut cables repaired and the stylus replaced.

So the robbery was very much on their minds when the heat wave hit a few days after they got back, when the temperature barely got below ninety, even at night, and was well over a hundred in the daytime, and it went on for damn near a week. Nobody could remember anything like it in San Francisco, and David had to keep reminding Danny, "When you feed the livestock"—handing him the cat food—"be sure their water dish is full. They can get dehydrated so easily in this kind of heat."

"Well, if they get that bad they can always drink out of the fish pond. They do already."

"Danny—"

"Dad—"

That night about ten he came out of the studio and stepped into the downstairs bathroom to shave. The heat reminded both him and Jane of their two months at the Mortimers', when they would make love in the afternoon, and earlier that day they had got a little turned on standing there in the backyard necking. Now he would shave and go on upstairs where she would be waiting for him in their bedroom wearing something sexy and they would make sure Danny was in bed and then— Both the cats were prowling around his feet as he rinsed off the last of the shave cream, looking into the mirror and thinking, So this is me, huh? Well, *Jane* loves it.

But when he looked down he saw that the cats were meowing their heads off and their fur was all fluffed out. When he tried to pick them up to put them out, they both squirmed free and hid behind the washing machine. "What the hell?"

He looked out through the glass panels of the back door, but could not see anything. He opened the door and looked out. Still nothing, but he could hear the bamboo wind chimes that Jane had hung from a corner of the roof clicking and tinkling away, and also a loud crunching sound, like something crisp being chewed. It took him a moment or two to realize something was wrong: the wind chimes were moving, but there wasn't any wind. The crunching, he figured, was one of the neighbor's cats who came into their yard periodically to rip off some of the dry cat food they left in bowls out on the patio for Girl and Geoffrey, along with their water dish. So as he walked out into the darkness he was already yelling, "All right, scoot! Get out of here!" and clapping his hands sharply.

But the crunching did not stop as it always did when the neighborhood cats scurried across the yard and over the fence, and the wind chimes kept rattling, and now something splashed once, loudly, in the fish pond. In the darkness of the corner where the cats' dishes sat, something big moved grayly. "What the hell?"

He backtracked up onto the deck and found the switch and turned on the light, and there they were, four raccoons, looking directly at him with eyes that were serenely unperturbed or threatened by his sudden appearance. One was *reclining* languorously on the roof, reaching down casually from time to time to pull on the string, rattling the chimes. Even now, looking straight at him, the raccoon reached down and calmly *twanged* the string. Two were in the fish pond, one holding a dripping paw to its mouth while the other looked away from him now to the water it was sitting in and made a scoop into the water and brought the paw up, dripping, to its mouth.

"They're after the fish!" David heard himself say aloud. He could only stand there with his mouth open. The closest one was not more than ten feet away, still busy scooping up pawfuls of the cats' food, dipping them into the water dish and then stuffing them into its mouth. Crunch.

There was something about the way they sat there, in his yard, as if they had appropriated it, as if it were theirs and they were putting up with his interruption with a disdainful patience. They infuriated him, with their bandit masks and their sharp little teeth and their sublimely unconcerned eyes. He yelled at them, "Get the hell out of here! Get out!" But they only continued to look at him. "There is no telling what some fool like this might do," they seemed to tell each other as they stopped crunching and eating and twanging just to be able to keep an eye on him better, but it was obvious that they did not feel in any way threatened by his presence.

He did not want to go down off the deck into the yard, for fear they might rush him. They couldn't hurt him very much, but there was no telling whether or not they were rabid or carried who knows what diseases, and he wasn't about to take any stupid chances. Clearly, he thought, they've come down from Mount Sutro, driven out of their usual hangouts by the drought and the heat. Raccoons don't have any salivary glands, so they have to soak their food or they can't eat it. So now here they are in my backyard. They've totally freaked out the cats, they're ripping off their food and God knows how many of the fish they've eaten or terrorized. He might have sent Danny out here to feed the cats, and in the dark the boy might have tripped over that one—and then what? What if they were rabid and one of them had bitten him?

He looked around for something to throw, but he could not see anything but a couple of potted plants sitting on the railing of the deck. He would have thrown them both if he could have been sure of hitting them, but he knew he would not be able to explain that to Jane. He saw a trowel and a weeding fork, which might have made good weapons, but they were too short. He went back inside and pulled the broom out of the closet, but after he hefted it he put it back. Too light. He picked up Danny's baseball bat and as he went back through the kitchen he saw the pans hanging against the wall and grabbed two of them. When he got to the back door, he looked out through the glass. One of them had actually come up onto the deck. God damn!

He opened the door and stomped his foot on the redwood planks. The raccoon stopped and turned toward him, looking as if it had not yet made up its mind whether to teach this fool a lesson or let it slide for now, but figuring it'd just better keep an eye on him.

Stomp! "Scoot! Get the hell out of here!"

Nothing.

He put the bat down where it would be in easy reach and grabbed the two pans by their handles. He clanged the bottoms of the pans together several times. The animal ducked and stayed crouched, low to the floor, looking as if it could just as easily rush him as make a run for it. Its eyes were cold and hard behind its mask. Absolutely disdainful. The two in the fish pond were now sitting upright, watching him closely. Even as they looked at him, one of them scooped down into the water again and brought its dripping paw up to its mouth.

He suddenly realized that he had lost track of the one on the roof. That one might have leaped down on him from behind. He looked behind him quickly, but to do that he had to take his eye off the one in front of him. He backed into the laundry-room door, where he stood for a moment in a frustrated rage. He did not want to tell Jane or Danny. He was going to teach these fucking animals a lesson, though. They could not come barging in like this, and just take over his backyard. This was his property. This was the city, damn it!

He wished he owned a shotgun. But if he did he would be shooting into McCormack's yard. Even if he didn't hit a person or a window, he would sure as hell leave pellets all over McCormack's wall.

The raccoon was now *shitting* on the deck. Oh, you son of a bitch! He had never felt anything like this rush of frustrated hatred before; he wanted to kill.

He went quickly to the hall closet next to the front door and got Danny's bow. By the time he got back to the laundry room, he had it strung and was hanging the quiver of steel-tipped arrows over his shoulder. The cats were peering out from behind the washing machine.

He looked out through the back door.

Nothing.

He opened the door and checked the roof line. He thought he saw a shape, but it did not move for a long time, so he walked out, grabbing the baseball bat as he went out to the middle of the deck where the one on the roof would not be able to jump him from behind.

The two were now climbing out of the fish pond, and even though they were not crouching behind it they were not giving him much of a target. He could not see the other two, though he could hear them moving around in the dark bushes just beyond the patio. They were out there, he knew, and these two, just sitting there, their heads showing above the tile of the fish pond. He threw the bat at them. It sailed about two feet over their heads. They did not even bother to duck, though they flinched and tensed up again when they heard the bat hit the ground some fifteen feet beyond them.

The smell of the raccoon shit made him even more furious. He wanted to kill them all, thinking it was a shame the burglar-alarm system couldn't have been rigged up with some sort of electrified fence or something so that now he could turn it on and chase them into it and watch them fry as they hit the wire. He pulled one of the arrows out of the quiver and set its notch in the string, holding its shaft on the grip with one finger. Now he was exulting that he had practiced this all summer. He knew he had a tendency to pull his bow hand up and slightly to the right just at the moment of release, so he aimed low, just at the tile rim of the fish pond. The bigger of the two animals was crouching now just beyond that rim. The arrow would hit it just below the eye, on the snout; the steel tip would pierce the bone of the skull and continue on through the brain and would probably come out again just below the back of its neck. The body of this one, as it looked at him, in full face, was so thick, so complicatedly thick, and the arrow would have to go through all those layers of skin and bone and organs. The other one was sitting now sideways, showing him its profile. He could feel his heart pounding with fierce joy, and felt somewhat annoyed to see how the steel tip of the arrow was moving around jerkily as he moved in his nervousness trying to take aim. He thought, You've got to keep your cool.

"No," he said aloud, putting down the bow. "This is crazy." He thought, I'm not Daniel Boone. I'm not some Indian defending his hogan against all the terrors of the wild. This is San Francisco. He was overreacting. He had been ripped off and had his territory invaded. And now these little animals with their cartoon burglar masks like the Beagle Boys, how could he want—so intensely, so madly—to kill them? They only wanted, they wanted things. They were hungry. Of course they reminded him of that burglar. Albert. Yet he too had had some kind of style, *strolling* into the lineup booth. He would have shat on the floor too, probably. Why not? What must it have been like to be Albert, tiptoeing into somebody's backyard and getting in through the back door? What must it have been like to hear your own breathing and your own heartbeat there in the space where *other people* had their lives? The two raccoons moved away now, back toward the darkness, as he laid down the bow. What must it be like to have your being like that, furtively, like an animal, and to come into a yard, like this one, and be suddenly confronted by a dog that would give you away by its barking? You would have to kill it. That must have been the hard part, killing the animals.

A NOTE ABOUT THE AUTHOR

Ron Loewinsohn was born in the Philippines and educated at the University of California at Berkeley and Harvard University. He is the author of five books of poetry and the co-editor of several anthologies, among them *The New American Poetry* and *A Geography of Poets*. He has been the recipient of numerous grants and awards including the Stone Award of the Academy of American Poets, a Woodrow Wilson Graduate Fellowship, a Danforth Graduate Fellowship, a Harvard Graduate Prize Fellowship, and a National Endowment for the Arts Fellowship. He lives in Berkeley, California.

A NOTE ON THE TYPE

The text of this book was set via computer-driven cathode-ray tube in Bembo, the well-known monotype face. The original cutting of Bembo was made by Francesco Griffo of Bologna only a few years after Columbus discovered America. It was named after Pietro Bembo, the celebrated Renaissance writer and humanist scholar who was made a cardinal and served as secretary to Pope Leo X. Sturdy, well-balanced, and finely proportioned, Bembo is a face of rare beauty. It is, at the same time, extremely legible in all of its sizes.

Composed by Crane Typesetting Service, Inc.,
Barnstable, Massachusetts

Printed and bound by The Haddon Craftsmen, Inc.,
Scranton, Pennsylvania

Typography and binding design
by Dorothy Schmiderer

PS3523 032 M3 1983
+Magnetic field(s+Loewinsohn, Ron.
0 00 02 0204778 3
MIDDLEBURY COLLEGE